AF358478

Inputs of Engineering Education towards Sustainability

Inputs of Engineering Education towards Sustainability

Guest Editors

Clara Viegas
Natércia Lima

Basel • Beijing • Wuhan • Barcelona • Belgrade • Novi Sad • Cluj • Manchester

Guest Editors

Clara Viegas
Physics Department
Polytechnic of Porto
Porto
Portugal

Natércia Lima
Physics Department
Polytechnic of Porto
Porto
Portugal

Editorial Office
MDPI AG
Grosspeteranlage 5
4052 Basel, Switzerland

This is a reprint of the Special Issue, published open access by the journal *Sustainability* (ISSN 2071-1050), freely accessible at: www.mdpi.com/journal/sustainability/special_issues/E5M06I8916.

For citation purposes, cite each article independently as indicated on the article page online and using the guide below:

Lastname, A.A.; Lastname, B.B. Article Title. *Journal Name* **Year**, *Volume Number*, Page Range.

ISBN 978-3-7258-2818-0 (Hbk)
ISBN 978-3-7258-2817-3 (PDF)
https://doi.org/10.3390/books978-3-7258-2817-3

Contents

About the Editors

Clara Viegas

Maria Clara Neves Cabral da Silva Moreira Viegas completed her PhD in Exact and Natural Sciences and Technology in 2010 at Universidade de Trás-os-Montes e Alto Douro, her master's in Mechanical Engineering in 1998 at Universidade do Porto—Faculdade de Engenharia, and her degree in Physics/Applied Maths in 1991 at Universidade do Porto—Faculdade de Ciências. Currently she is an adjunct teacher at Instituto Politécnico do Porto Instituto Superior de Engenharia do Porto, a full researcher at the CIETI at Instituto Superior de Engenharia do Porto (P.Porto), and co-director of the degree Licenciatura Engenharia Biomédica at Instituto Politécnico do Porto Instituto Superior de Engenharia do Porto. She has published 25 articles in journals, 21 book chapters, one book, and 54 conference papers. She has co-organized five international events. She has served as an arguer and main arguer on seven PhD juries, as the track chair at various International Conferences, and as a guest editor of Special Issues in research journals. She works in the area of engineering education with emphasis on physical sciences. Her research interests are active learning; assessment and feedback; competence development; learning and teaching strategies; learning motivation; multimodal narratives; physics' didactics; professional development; experimental work; remote labs; sustainable development; teacher mediation in classrooms; and teaching introductory physics.

Natércia Lima

Natércia Lima was born in Porto, Portugal, on December 11, 1965. She received her bachelor's degree in Physics/Applied Mathematics and her M.Sc. degree in Mechanical Engineering from the University of Porto, Portugal, in 1989 and 1998, respectively. In 2020, she received her PhD (cum laude) in Education in the Knowledge Society (in the area of engineering education) from the University of Salamanca.

She is a professor at the Polytechnic of Porto—School of Engineering (ISEP) since 1993 and currently holds a position as an adjunct professor. She has been a researcher at the CIETI—LABORIS—Center for Innovation in Engineering and Industrial Technology since 2014. She has authored or co-authored over 45 publications, including chapter books and conference and international scientific journal papers, with a referee process. She has participated in two international research projects. She has been a member of several program scientific committees of international conferences and a member of organizing committees, including the local chair of national and international conferences/symposiums. She has collaborated as a reviewer of several international journals and conferences in engineering education topics. Her research interests include engineering education (with special attention to the use of virtual and remote labs in students' learning outcomes), students' active learning, and teachers' professional development.

She has also been a member of the ISEP's Scientific Technical Council (2009–2014), a subdirector of the Master in Instrumentation and Metrology Engineering (2012-2016), and a member of the ISEP's Physics Department Management Board (subdirector: 2002, 2007–2010; director: 2003–2005; 2010–2012).

She is currently (since May 2022) the subdirector of the Master in Biomedical Engineering.

Preface

Education is the key to progress in every society. Educational institutions are places where people can furnish their ideas and develop corroborated arguments to discuss each of the sustainable development goals (United Nations SDGs). Educational institutions are therefore a powerful piece in this puzzle, where awareness, passion, and strategies may flourish. The awareness of the importance of sustainability issues is the first step towards a solution. This role is of particular importance in higher education institutions, where knowledge, skills, and values are crucial in directly tackling issues like climate change, energy and resource overuse, etc. By simply promoting actions, school debates, or stimulating the creation of academic groups that address these issues informally, sustainability can be integrated into different aspects, if not all, of education and training in a process of lifelong learning.

One of the major contributors to support sustainability concerns is engineering. Teachers have a key role in this process, namely in the development of students' awareness of sustainability issues. This is especially true in technical and applied fields such as engineering. Teachers can adopt approaches and practices that encourage the training of future engineers with a systemic and responsible view of the environmental and social impacts of their activities. They may address this problem in several ways.

Every teacher's action, simple as it may be, can positively impact students' behavior in class, in their homes, and in their communities. These efforts replicated by other teachers all over the world, who may follow similar practices, will certainly be a step forward. Even though they represent small contributions, the increase in conscious awareness of the problem is very important per se.

All, in different ways, attempt to demonstrate in practice how it is possible to apply sustainability to small daily choices, setting examples for students. These practices can transform engineering education, training professionals who are more aware and capable of facing global environmental challenges and actively contributing to the transition towards a more sustainable society.

The works received in this Special Issue clearly illustrate how teaching may contribute to sustainability on a small scale but hopefully with a widespread scale. These approaches aim to help students integrate sustainability into the way they think and work, building a strong foundation of values and practices to become engineers committed to a sustainable future.

From the 16 papers submitted, 10 were accepted and published. Congratulations to all authors who shared their contributions to this Special Issue and helped us reflect about how engineering education can play its role in sustainability development from such different perspectives.

Clara Viegas and Natércia Lima
Guest Editors

 *sustainability*

Article

Adding Machine-Learning Functionality to Real Equipment for Water Preservation: An Evaluation Case Study in Higher Education

Maria Kondoyanni [1], Dimitrios Loukatos [1], Konstantinos G. Arvanitis [1,*], Kalliopi-Argyri Lygkoura [1], Eleni Symeonaki [2,*] and Chrysanthos Maraveas [1]

[1] Department of Natural Resources Management and Agricultural Engineering, Agricultural University of Athens, 75 Iera Odos Str., 11855 Athens, Greece; mkondoyanni@aua.gr (M.K.); dlouka@aua.gr (D.L.); stud616018@aua.gr (K.-A.L.); maraveas@aua.gr (C.M.)

[2] Department of Industrial Design and Production Engineering, University of West Attica, Thivon 250 and P. Ralli, 12244 Egaleo, Greece

* Correspondence: karvan@aua.gr (K.G.A.); esimeon@uniwa.gr (E.S.); Tel.: +30-210-5294-109 (K.G.A.)

Abstract: Considering that the fusion of education and technology has delivered encouraging outcomes, things are becoming more challenging for higher education as students seek experiences that bridge the gap between theory and their future professional roles. Giving priority to the above issue, this study presents methods and results from activities assisting engineering students to utilize recent machine-learning techniques for tackling the challenge of water resource preservation. Cost-effective, innovative hardware and software components were incorporated for monitoring the proper operation of the corresponding agricultural equipment (such as electric pumps or water taps), and suitable educational activities were developed involving students of agricultural engineering. According to the evaluation part of the study being presented, the implementation of a machine-learning system with sufficient performance is feasible, while the outcomes derived from its educational application are significant, as they acquaint engineering students with emerging technologies entering the scene and improve their capacity for innovation and cooperation. The study demonstrates how emerging technologies, such as IoT, ML, and the newest edge-AI techniques can be utilized in the agricultural industry for the development of sustainable agricultural practices. This aims to preserve natural resources such as water, increase productivity, and create new jobs for technologically efficient personnel.

Keywords: internet of things; machine learning; smart sensors; fault detection; embedded systems; smart agriculture; water preservation; sustainability; educational practices; higher education

Citation: Kondoyanni, M.; Loukatos, D.; Arvanitis, K.G.; Lygkoura, K.-A.; Symeonaki, E.; Maraveas, C. Adding Machine-Learning Functionality to Real Equipment for Water Preservation: An Evaluation Case Study in Higher Education. *Sustainability* **2024**, *16*, 3261. https://doi.org/10.3390/su16083261

Academic Editors: Clara Viegas and Natércia Lima

Received: 29 February 2024
Revised: 29 March 2024
Accepted: 9 April 2024
Published: 13 April 2024

1. Introduction

Nowadays, innovative technologies have been incorporated into the curricula of schools, delivering promising educational outcomes, particularly in the STEM (Science, Technology, Engineering, and Mathematics) framework. Although STEM education has many benefits for young students, the same courses and principles cannot apply to higher education without proper adaptation, as higher-education students, having already acquired several basic knowledge sets and skills, seek advanced learning experiences [1]. Quite a few innovative solutions are available that support laboratory-level trials and create interesting activities that encourage the students to take part in all stages of the development of real-world applications. Indeed, considerable work is done, involving microcontrollers, pairing electronics [2], and small-scale inexpensive systems (i.e., robotic devices), usually through applying PBL (Project Based Learning) approaches [3], in the context of K-12 education [4,5] and higher education [6]. The inclusion of the STEM model into today's educational methods is challenging for a variety of reasons [7]. In the case of university students, more effort should be made, as the students ask for more complex systems and learning experiences that are equal to real-world conditions.

There is apparently a gap between university and industry, regarding appropriate engineering capabilities [8,9], which is easier to bridge by developing pure software, using open platforms [10]. Additionally, software platforms are highly beneficial for facilitating the development of systems and programming controllers of industrial specifications [11–13], including the ones applied in the agricultural sector [14]. It is worth mentioning that, quite frequently, students are not being given the opportunity to experiment with the full set of diverse settings of actual equipment during the classes, which mainly operates as a "black box" system [15]. This situation, inherited by the secondary education, is experienced in the engineering university department and is further magnified due to the intensification of the students' technical curricula. Additionally, the outdated equipment being used on several occasions during the laboratory classes cannot always be computerized (i.e., with modern interfaces for communication, monitoring, and control) or it is merely computerized, solely with the assistance of properly trained personnel.

These facts indicate the necessity and the difficulty of incorporating advanced technologies into the universities' courses [16], not only in theory but also in practice, via enriching laboratory activities with practical experiences and, thus, aiming to train students so as to develop skills for tackling real-world problems. In this regard, the inclusion of innovative technologies and hands-on activities in higher education will prepare students to adapt easily to the continuously developing industry and to gain valuable knowledge and skills that will be utilized in the future to improve the planet. These actions need to be carried out taking into consideration the fostering of sustainable development in order to ensure environmental protection and preservation of natural resources [17].

Indeed, sustainability serves as a main pillar for the green transition, it entails fulfilling current needs while ensuring future generations can meet their own needs without compromises [18]. Hence, sustainable practices intend to decrease the negative impact of human activities on the environment, society, and the economy [19]. In this context, entrepreneurship should enable the development of new technologies and business models that provide fresh products, and services that create value by addressing sustainability challenges, while also contributing to economic growth and job creation [20]. Sustainability and education are intricately linked concepts that play a pivotal role in shaping a sustainable future for our planet. The idea of Sustainable Education (SE) involves seeking lasting solutions to environmental, social, and economic challenges through educational means [21]. This concept calls upon both formal and informal education sectors to engage proactively in developing programs that enhance quality of life, promote empowerment, and recognize the interconnectedness of economic, social, and environmental aspects [22,23]. On the other hand, Education for sustainability refers to the integration of sustainability principles into the curriculum and educational experiences. It focuses on teaching students about concepts, values, and skills related to sustainability, enabling them to understand and address complex global challenges, such as climate change, biodiversity depletion, and social inequality [24]. It emphasizes the content of education, incorporating topics related to environmental conservation, social justice, and economic viability into various subjects. Students are encouraged to explore real-world issues, develop critical thinking skills, and participate in projects that promote sustainable practices in society. The goal is to create environmentally and socially conscious citizens who can contribute to building a more sustainable future.

From the perspective of agriculture, dealing with the problem of water depletion, according to sustainable policies, is of particular concern since the agricultural sector is the main water consumer on Earth. In fact, water is an important input for agricultural production and holds significance in food security, as global irrigation for agricultural production comprises 70% of clean water use [25]. Water pumps and faucets play a vital role in irrigation operations, as they are primarily employed to transport substantial volumes of water from their respective sources to the fields. Furthermore, water pumps can undergo damage due to various factors, including insufficient provision of water from the origin, inefficient power, or the circulation of contaminated water. To avoid damage to the components of these pumps, it is crucial to observe their operational status and act in the event of a malfunction. The digitalization of agriculture appears to be a promising opportunity for monitoring and automating agricultural operations through cutting-edge technology, such as the internet of things (IoT) and machine learning (ML) [26,27].

For the aforementioned reasons, the case described in this paper focused on addressing water depletion issues and equipment maintenance. This approach aimed to familiarize students with the principles of sustainability, encouraging them to comprehend the intricate interplay between human activities, environmental concerns, and sustainable solutions. Above this, the contribution of this work is to emphasize the viability for integrating software and hardware components to provide, apart from technical outcomes, considerable educational outcomes regarding the issue of strengthening the benefits of real equipment for water preservation. These components, without being expensive, can assist in creating effective instruments for maximizing the educational benefits for the students, while they are called to tackle problems related to the water preservation purposes. The experiments carried out were dedicated to sustainability, emphasizing the preservation of water and the maintenance of equipment. This method strives to achieve dual advantages for participants: acquiring additional technical knowledge while also putting sustainable applications into practice. In this context, the experimental arrangements described herein are trying to highlight the potential benefits—from technical and educational perspectives, with more focus on the latter—of modern technological advancements like the ML and the IoT, aiming at the delivery of cheap devices capable of making smart in situ decisions and reporting the results to the interested parties accordingly, instead of relying upon complicated, non-cost-effective, centralized infrastructures.

In more detail, the first step of this research, in the direction of the on-device intelligence deployment technique [28], was the development of a classification model executed on a microcontroller attached to a commercial faucet along with a flow sensor so as to determine water-consumption profiles and alert the user about them [29]. To further benefit from this deployment technique, a machine-learning model was developed for classifying and diagnosing possible motor defects in a water pump, using vibration data from an accelerometer to achieve a more comprehensive and precise perspective on the factors influencing the system [30]. These preexisting works, apart from introducing technical innovations, provide fertile ground, from an educational perspective, for pedagogical setup descriptions, experimentation, and evaluation reports that are among the main subjects of the study being presented. Toward this direction, the applications' implementation process and overall utilization experience are also evaluated through questionnaires, suitable for the specific target groups of undergraduate and postgraduate university students. The results indicated that valuable hard skills and soft skills were acquired (and, thus, reported) by the students who participated, thereby making them better prepared for their roles in a rapidly changing era.

Subsequently to the introduction in Section 1, the paper is laid out as follows. Section 2 identifies the main motivations and challenges of this work. Section 3 provides an overview of the educational arrangements as well as some facts about the functionality and the selection of the components. In Section 4, the design of the system is presented and some interesting details regarding its implementation are highlighted. Section 5 concentrates on evaluation of the results and discussion of some insights derived from the findings of this work. Finally, Section 6 of the paper presents key conclusions derived from this work and outlines potential directions for future research.

2. Related Work and Rationale

In terms of agricultural engineering education, in most cases, universities primarily prioritize enhancing the performance of particular implementations from a technical standpoint, often without placing emphasis on the basis of educational practices or the social impact [31]. It is important to highlight that students in agricultural engineering show a clear preference for teaching methods based on experiential learning, which equips them with skills for innovation and creation [32,33]. The field of digital agriculture includes not only agriculture, but also engineering and computing, and, thus, it is challenging to find experts that are fully conversant with these aspects simultaneously [33]. Moreover, education for sustainable development [34] is not included in the activities of universities, although it is a key element in the Agenda for Sustainable Development, driving the fulfillment of all the Sustainable Development Goals (SDGs) [35]. The ESD stands for the inclusion of sustainability issues, i.e., protection and conservation of natural resources, climate change, and sustainable exploitation/consumption, in teaching and learning [34]. In this regard, preparing well-trained professionals should include equipping them with the knowledge, skills, and values that will empower them to contribute to the creation of a more sustainable world and enhance wellbeing as well as socio-economic growth, along with conserving natural resources. The importance of introducing the sustainability concept in higher education has been identified by researchers. As [36,37] indicate, in order to emerge as leaders and catalysts for change in sustainability, higher education institutions must prioritize understanding and addressing the needs of both present and future generations. This involves equipping professionals, well-versed in Sustainable Development (SD), to effectively educate individuals of all ages and guide them in transitioning to sustainable societal patterns. To achieve this goal, it is crucial for university leaders, faculty, and students to be empowered to introduce Sustainable Development into all aspects of their institutions, including courses, curricula, and various activities. Recognizing the importance of multidisciplinary and transdisciplinary approaches in teaching, research, and community outreach is essential for expediting the necessary societal transition toward sustainable development.

The literature also shows that the incorporation and utilization of technologies in education systems, overall, are not advancing as indicated by the digital and 2030 agendas [38] due to technological, pedagogical, and organizational inefficiencies. Many of the new technologies having a strong impact in modern life are not well incorporated yet into the higher education curricula, that remain more theoretical than practical, while flexibility and multidisciplinarity are required. A notable case of such technologies is the field of artificial intelligence (AI) known as machine learning (ML), that has many practical applications offering solutions to several critical problems, and, thus, could be making it a prime example motivating for integration into educational practices.

Apart from the more conventional educational approaches dealing with engineering with electromechanical [39,40] and basic IoT [41] solutions applied in modern industry and agriculture, various educational projects aim to improve individuals' AI literacy. According to recent research [42], preliminary courses on AI are offered at various educational levels, from elementary school [43] and secondary education [44,45] to higher education [46,47]. Nevertheless, most of these approaches are software-based paradigm and they are not well linked with real-world problem solutions that fully exploit the engineering spectrum. Unfortunately, it is hard to find research works combining education on machine learning with impact on sustainability and offering at the same time real-world performance applications experiences.

Some works may be found, dealing with sustainability or with machine learning, although not satisfactorily covering both issues. machine learning models possess the ability to learn and adjust according to the problem, whereas traditional programming alternatives are constrained since those implementing them are expected to already understand the intricacies of the system for which the solution is being customized [48]. In recent years, the accessibility of extensive volumes of data and information has enabled more fast and accurate ML models [49]. The swift growth in available data, facilitated by improved sensor inventions, has significantly elevated the significance of machine learning, transforming it into a potent instrument for numerous applications between various disciplines. Indeed, fresh hardware and software tools have recently appeared, allowing for fast deployment of applications in the area, and it is worth these tools to be efficiently and creatively utilized by higher education professionals.

This work initially aims to bridge the aforementioned gap and proceeds further to achieve mutual benefits from the educational and technological context. In greater detail, the combined goal of this work is to facilitate the communication of innovative technology practices to agricultural engineering university students, based on the development of a final product, in order to become more efficient in their careers, and simultaneously to make the students aware of critical sustainability issues. In this regard, the activities being proposed need to be oriented towards covering all these types of challenges. Traditionally, through technology, several solutions had to be found to tackle intense problems, such as the depletion of natural resources or the increased nutritional needs of humans, while modern disciplines like IoT, automation control, artificial intelligence, and networking were amongst the most promising instruments of the abovementioned efforts. Therefore, the emerging advance in the area is further increasing the need for well-trained students and future professionals involved in developing, parameterizing, and maintaining the relevant systems.

Going deeper, it emphasizes the feasibility of developing economical systems of realistic dimensions, which is achieved due to the presence of user-friendly programming software, which can be either textual or visual, streamlining the entire approach. Indeed, according to the study report of 2021 of the European Commission (EU) [50], the role of open-source software and hardware is paramount for facilitating the digital transformation and fostering the improvement of societies. Additionally, from an educational perspective, the proposed processes, which are mainly oriented (but not limited) to agricultural engineering students, are utilized for better delivering the essentials of machine-learning techniques and various hardware, software, and networking principles as well as to raise awareness of sustainability issues and ways to contribute to more sustainable agricultural production practices. The demanding collaboration needed for the completion of the suggested system also offers the essential setting to reinforce several cooperative and organizational skills. Apart from that, the exploitation of retired or remaining/unused components is a good option, as they are inexpensive for the creation of educational scenarios, and they align with the common guidelines for sustainability and circular economy that modern communities are encouraged to adhere to [51]. In greater detail, during laboratory lessons experiences, university students of little technological background, were assisted to clarify cutting-edge technologies, and to bridge the gap between small-sized educational constructions and real-size systems. The experiments conducted have a very clear technological description in order to be easily reproduced by other teams of researchers/educators, but they are also strongly oriented towards sustainability, as they are dealing with subjects that intrinsically exist in the sustainability context, such as water preservation and pump equipment maintenance challenges.

To that end, this paper takes into consideration the material provided by two studies that use machine learning techniques for developing detection systems in order to address typical irrigation network problems. The first one introduces a water-misuse alert system [29], while the second one utilizes a classification model to detect water pump malfunctions in agricultural premises [30]. This article, except from providing a brief technical overview, is trying to explain how the latter systems can be transformed into effective educational instruments, suitable for serving the priorities of an agricultural engineering laboratory. It is an attempt to delve deeper into the integration of ML in the field of engineering from a scientific and educational standpoint, providing university students with the opportunity to combine hands-on methods and create smart agriculture solutions, often called "the future of the digitalization of farming" and "the driver of sustainable development".

3. Methods and Materials

Section 3.1 delves into the field of education for agriculture, defining the pedagogical goals and framework for acquiring both technical (hard) as well as interpersonal (soft) skills. Section 3.2 offers a concise summary of the enhanced farming systems and the rationale behind their design, to aid the understanding of the article.

3.1. Pedagogical Approach

From a pedagogical point of view, it is considered that the development of real proto-type systems, for example a smart-agriculture application, will function as a crucial tool for problem-solving and aid in the integration of various disciplinary domains. Note that an indicative review of the STEM educational directions along with the related trends, can be found in [52–54] and the references within them, whereas the advantages from the synergy of incorporating STEM practices with agricultural are shared and significant [55,56].

In this regard, aiming for a more effective education in agricultural engineering, a water usage alert system was developed, and a retired water pump was exploited in order to be transformed into educational instruments. Below, the fundamental goals of the suggested approach, concerning the acquisition of skills, are referred:

- Enhanced comprehension of machine learning basics,
- Enhanced comprehension of networking basics,
- Enhanced comprehension of embedded systems basics,
- Improved ability of students to model and solve real-world problems,
- Equipment of learners with knowledge, capabilities, and values that contribute to sustainable development.

Furthermore, to cultivate better pedagogical results of the students' training in this approach, one priority was the development of several soft skills, including:

- Enhancing the students' communication and team-working skills,
- Enhancing students' confidence of their professors' efficacy,
- Assisting students' self-confidence to accomplish a project based on given instructions.

According to the abovementioned analysis of the expected outcomes, it is anticipated that students participating in these activities will demonstrate enhanced learning potential, improved skill development, and improved learning capacity for innovative technologies fostering sustainability. To evaluate the influence of the suggested arrangements on the attitudes of the students were recorded anonymously and voluntarily using five-point Likert questionnaires.

Over the span of the 10-month duration of the core activities related to machine learn-ing, the persons participating in were: agricultural engineering professors (normally, one professor or two for each lesson activity), students working on their final thesis, students undertaking internships, and students involved in the curricular lesson activities. The mix of courses that the students attended during the semesters were: "Applications of Informatics in Agriculture", "Measurements and Sensors", "Electronics and Micropro-cessors", "Automatic Control Processes", and "Applications of Artificial Intelligence in Agriculture". Most students were in the age range of 20 to 26 years old. A team formation scheme was essential, aiming to assemble each group with members that had different but complementary capabilities, to some extent in accordance with the principles outlined in [57].

Challenge-Based Learning (CBL) provides an efficient framework for learning while solving real-world challenges, as it is an innovative teaching methodology that engages stu-dents to resolve real-world challenges while applying the knowledge they acquired during their professional training. Participants are encouraged to develop increasing interest for the subjects to be studied motivated by the significance of the problems to be addressed and their impact on society and well-being. Indeed, the CBL model has been applied to a large extend in higher education for groups of undergraduate students [58–60], and postgraduate students [61] and the results were positive, showing that the participants came up with innovative ideas to resolve challenges and improved skills and competencies. The benefits of the PBL [3] and CL [62] approaches, in terms of practicality and methodology, can be combined [63] and reinforced by the CBL technique to maximize the educational outcomes. As stated by Sukacke et al. [64] the implementation of active learning methodologies in education, such as PBL and more recently CBL, has become the new norm, especially in engineering universities, preparing future engineers for their professional careers.

The above philosophy is followed by activities being discussed, during which the instructors had the role to encourage and inspire the participants and to supervise the entire process, while the students in their teams, shared ideas and collaborated seeking for necessary information, comprehended techniques, conducted experiments, and executed challenging tasks of progressive difficulty. The students with greater experience served as mentors for the less experienced ones [65], thus assisting their professors.

Consequently, the primary challenges in implementing AI-based systems revolve around the absence of standardized programming practices and the insufficient multidisciplinary background knowledge of both trainers and trainees [66], thus selecting easy-to-use electronic parts as well as popular programming tools is needed. For the above reason, the activity being presented upgrades a water pump of convincing size and a faucet being used in real-world applications, while remaining plain and utilizing tools and components possessing these attributes.

3.2. Functionality Overview and Component Selection

This section provides, in brief, the overall functionality of both systems and the components selection for their development, while it additionally reports their role to be capable of executing important work related to water usage and pump functionality.

The aim of scenario A was the development of a system that can intercept and characterize water usage events. The water alert system included sensor nodes positioned at the place (edge points) where water is being consumed, along with appropriate sink/gateway node(s) to gather the reports transmitted by the peripheral nodes. The edge nodes, apart from collecting time series data corresponding to events that contained information about water usage, also possessed the intelligence to classify these events into categories of rational or irrational water consumption without depending on external entities. The user could monitor the operation of the entire system using portable devices (e.g., tablet, smartphone, or laptop) through traditional connectivity options.

In the case of the malfunction detection system installed on the water pump (scenario B), some common malfunctions were emulated for comparison with the normal operation of the water pump. Motion sensor data was recorded for four classes, which correspond to normal operation, and three cases of malfunction (inlet choke, outlet choke, and air intake) in addition to the fifth class of data with the engine switched off to simulate cases of inactivity. Therefore, a dataset containing five distinct classes was generated and a neural network model was developed that had the capability to identify each class. Additionally, a webpage was created to offer information to the user about the operation condition. More specifically, the hardware components being utilized were an AC centrifugal water pump (which was retired equipment), a water tank of a capacity of 50 lt, placed on a custom base and connected to the water pump via $\frac{3}{4}$ plastic hydraulic tubes interceded by plastic valves within each tube. These valves were quite significant components as they enabled the simulation of possible malfunctions. Moreover, readily available, well-documented, and cost-effective off-the-shelf hardware modules were employed.

Both systems incorporate an Arduino Nano 33 BLE Sense (Arduino, Turin, Italy) [67], which is a microcontroller board equipped with a robust processor that provides the ability to create bigger programs when compared to an Arduino Uno (Arduino, Turin, Italy) [68], with a 32 times bigger flash memory, and 128 times bigger RAM. In addition, a unit, based on the ESP8266 (Espressif Systems, Shanghai, China) [69] chip, for Wi-Fi connectivity options investigation, was used.

The Arduino Nano 33 BLE Sense device needed to be programmed in a way to:

- record and upload flow sensor data, specifically the interrupt signals corresponding to the pulses of the rotor rolling in the water flow sensor (scenario A),
- record and upload motion sensor data, specifically through its built-in accelerometer (scenario B),
- enable the essential networking connectivity (scenarios A & B),
- run the ML models (scenarios A & B),
- modular deployment with provision for implementations of diverse complexity.

The above prerequisites were fulfilled through the Arduino IDE (1.8.16) [70] programming environment, that was the most promising choice.

Regarding the machine learning part corresponding to both systems, training, and integration of each artificial neural network (ANN) [71] model into the software of the microcontroller was needed. Typically, the structure of an ANN model features a single input layer and some interconnected hidden layers, along with an output layer to provide the results. The Edge Impulse (EI) cloud environment was the platform chosen for the development of the ML models, as it is a straightforward and effective platform for developing ML models (encompassing training and extraction/compilation processes) tailored for edge devices [72]. EI accommodates a variety of development boards, including the Arduino Nano 33 BLE Sense, facilitating the immediate recording along with uploading the samples needed for the dataset. After the training, it also allows to directly deploy the model to the development board.

To make the procedures easily comprehensible for the students, the following deployment strategy was implemented:

- Installing a flow sensor on a typical water faucet (scenario A);
- Creating the basic water pump—water reservoir system plus the necessary valves emulating specific disturbances and installing an accelerometer sensor on it (scenario B);
- Connecting the sensors to a microcontroller so as to inspect and collect the readings of scenarios A and B;
- Training the ML model and installing it on an ML-capable microcontroller (e.g., the Arduino Nano Sense);
- Building an elementary networking functionality for easy inspecting the smart decision results.

The design and execution of the project were purposely made modular to clearly define the specific functions of each component. This was aligned with the university curriculum and helped agricultural engineering students to better understand modern technologies. Despite using relatively simple components, a few challenges arose, triggering the interest of students, particularly due to the real-scale nature of the systems being proposed.

4. Design and Implementation Details

Section 4 is devoted to describing the crucial implementation details and challenges of the presented farm systems. In particular, Section 4.1 furnishes technical details pertaining to hardware and software issues while Section 4.2 covers information regarding neural network training. Section 4.3 specifies the details of on-device integration, and finally Section 4.4 outlines the hardware and software for the system at the end-user.

4.1. Description of the Basic System

The proposed implementation, regarding the water alert system, was built around an Arduino Nano BLE unit, being the coordinator of the main functions. For measuring the water flow, a Hall effect meter sensor (YF-S201 model) [73] was employed, capable of detecting changes in the flow of water as it passes through and rotates the rotor. This system is intended to be fixed close to a tap/faucet, so as the flow sensor to be connected in series with the water supply pipes, as depicted in the design overview of the system in Figure 1a. The second system that is being discussed involved a water pump that was designed to operate as a closed system, meaning that water was drawn from a 50-L tank and recycled back into the tank. Plastic tubes were used to connect the inlet and outlet of the water pump to the tank, intercepting by plastic valves to control the flow of water in each tube. An additional tube was attached vertically to the inlet tube of the water pump and had an open end that allowed air to enter based on the position of a valve. The Arduino Nano, equipped with an integrated accelerometer, was positioned on the side of the water pump to identify and differentiate the vibration patterns for each dataset. (Figure 1b).

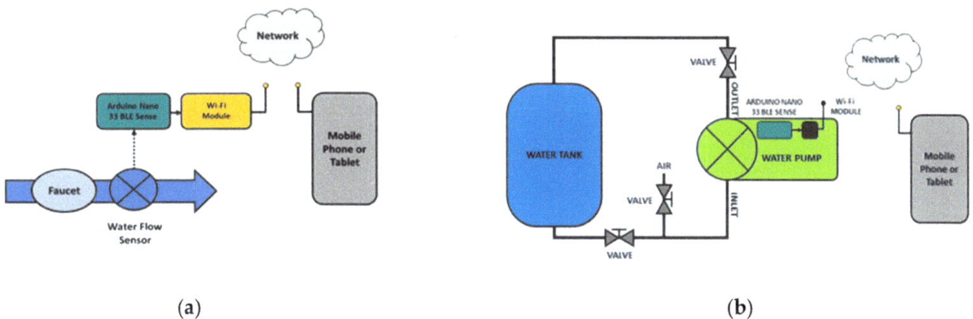

Figure 1. (**a**) Design overview of the water misuse system added in series to a standard faucet (system A); (**b**) Overview of the design of the malfunction detection system for water pump operation (system B).

During the implementation stage, to acquire the necessary data for training the model, a connection was established between the Arduino and a computer, via serial interface, thus enabling the straightforward recording, and uploading samples to the designated Edge Impulse project. This connection was also used for the compilation and uploading of the trained model to the Arduino Nano BLE, as well as to monitor the model's performance. Last but not least, the Arduino Nano BLE was connected to an ESP8266-based radio [74] to enable remote network connectivity, via a Wi-Fi link.

4.2. Neural Network Training

An artificial neural network (ANN) is a kind of machine learning model that is designed to emulate the configuration and functionality of the human brain, comprising interconnected nodes, or "neurons", responsible for processing and transmitting information [71]. The fundamental phases of a machine-learning model training and deployment can be summarized into 4 stages as shown in Figure 2. The initial phase involves acquiring data, a critical step that enables the training of an ANN model, which will impair the entire system with machine learning capabilities. This includes formatting the data, as well as splitting it into training and testing sets. The second and third steps detail the training process. During these stages, the suitable parameters are chosen to train the model by utilizing the prepared data and any required improvements are made, such as the learning rate and number of layers. Then using the testing data, the model evaluation is performed. This stage is crucial to determine whether the model requires modifications and to obtain an initial assessment of its accuracy. In the fourth step, the trained ANN model is integrated into the edge device (e.g., microcontroller) to enhance the system's functionality. The deployment of the optimized model to the edge device involves converting the model into an appropriate format and integrating it into the device's software or firmware. The deployed model should be tested to ensure its correct and accurate operation on the edge device.

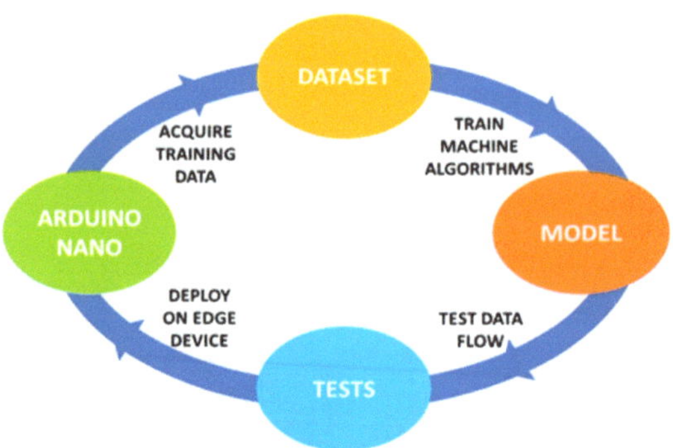

Figure 2. The necessary stages for the machine-learning training of the systems described and deployment on an edge device.

Firstly, in order to train a neural-network model, it is necessary to gather an adequate quantity of data for each specific class. The dataset for the water-pump malfunction-detection system included 5 classes: one for data corresponding to normal operation, three for simulated failures, and another for data categorized as noise. The length of the sample was five minutes or more, which is sufficient for this work, given its primary educational purpose. The gathered data must be divided into a training dataset, employed for training the neural network, and remaining data should be set aside to test the model's efficiency. When the data collected is automatically uploaded to the training set, it is recommended to allocate approximately 20% of them to the testing dataset. Nevertheless, this percentage can differ slightly since the total number of samples may not be divided accordingly. For these models, the split was conducted at a ratio of 78% to 22%, which did not have an impact on the model's overall accuracy or effectiveness.

Once the requisite data has been collected and segmented, the subsequent step involves designing and training the model. This phase entails the incorporation of a processing block to modify the data and of a learning block that facilitates the selection of the specific neural network to be trained. The information was gathered using the built-in accelerometer of an Arduino Nano. To process this type of data, the appropriate block, "Spectral Analysis", was chosen since it can analyze continuous motion, such as accelerometer data, and extract the signal's frequency and power characteristics over time. In the learning block, we utilized a classification neural network library implemented with Keras. This library is equipped to learn patterns from input data and implement them to new data. This library is particularly well-suited for recognizing audio or categorizing movement, with the latter being the primary focus of this experiment. Moreover, the window size was configured to be 2000 ms (equivalent to 2 s), in accordance with the profiles input into the training system, taking into account the duration of the phenomenon. Similarly, the window increment was established at 80 ms, and the frequency was set at 100 Hz.

Yet for the water pump malfunction detection system, the processing block produced 33 features, which are then imported as the input layer in the training procedure. The intermediate layers comprised 10 and 5 neurons, respectively. Thirty training cycles were set, and the output layer encompassed the 5 classes. Following the training process, Edge Impulse can store the best performing model in the Quantized (int8) version, suitable for the Arduino hardware platform.

The same approach was followed for the development of the ANN model for the water usage alert system. The data related to water flow was transferred to the Edge Impulse cloud platform and then manually labeled before being automatically divided into training and testing data. To train the ANN model, a window size of 200 s was set based on how long someone uses a faucet, with a window increase of 1 s and a frequency of 1 Hz. Moreover, the "Raw Data" was chosen as the suitable processing block, along with "Classification (Keras)" as the learning block for the ANN. This allows for the original data to be used without any additional processing, retaining as many attributes of the original data as feasible. Thus, the neural network had an input layer with 200 features, two hidden layers with 20 and 10 neurons, respectively, and an output layer with three classes, namely NU, WL, and WW. The model was saved in the quantized version, that occupies 1.9 KB of RAM and 22.5 KB of flash memory, allowing it to be uploaded to the Arduino and run in real time. Indeed, the Edge Impulse platform allows to efficiently experiment with different settings in order to keep the best-performing model. The latter model can then be downloaded, as code, from the Edge Impulse platform that encompasses the library and sketches to be compiled and uploaded to the microcontroller using the Arduino IDE environment.

4.3. On device Integration Details

In the smart faucet case, the flow sensor and the ESP-01 radio module are powered from the Arduino Nano Sense BLE unit. On the latter device, interrupt signals are enabled so as to intercept the flow pulses, while the communication with the radio module is established through its hardware serial interface. Figure 3a depicts details of the prototype smart faucet system, while Figure 3b highlights the main electronic components of the water-pump malfunction-detection system.

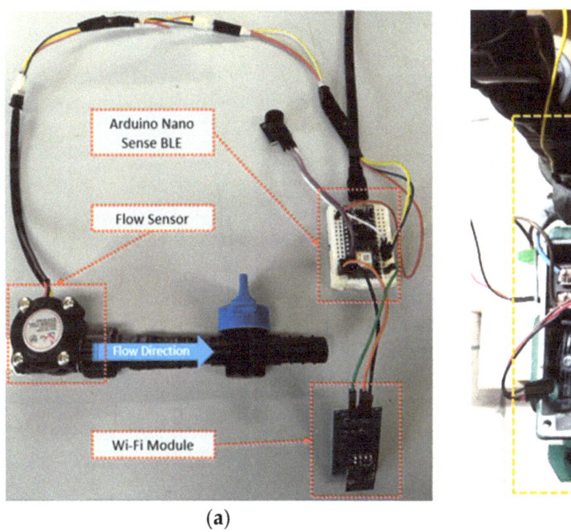

(a)   (b)

Figure 3. (**a**) Details of the prototype smart faucet system; (**b**) Details of the prototype water-pump operation malfunction-detection system.

In both systems, for simplicity and safety reasons, the power supply is done via the USB port of Arduino. Nevertheless, the flexible powering options of the main microcontroller were exploited to familiarize the students with battery-based variants, offering improved autonomy using a Li-ion battery of 18,650 type.

To enable real-time alert generation reflecting either pump malfunction or water misuse event classification, the Edge Impulse platform generated code in the form of an Arduino library as indicatively shown in Figure 4a, which could be used with the Arduino Nano 33 BLE Sense board. This flexible option helps to integrate the native machine-learning model with supplementary algorithms. Figure 4b depicts a screenshot of the Arduino IDE environment throughout the time of programming.

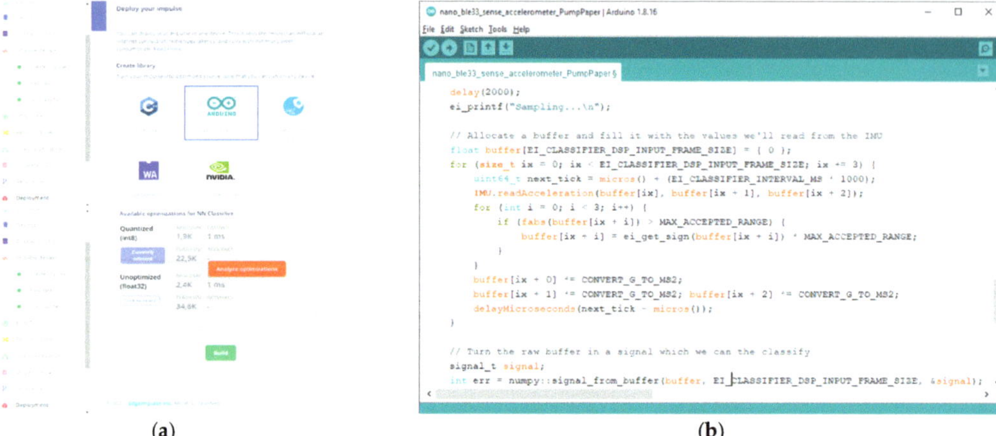

(a)   (b)

Figure 4. (**a**) Indicative neural-network model deployment options via the Edge Impulse platform; (**b**) Indicative code including the trained model and the necessary modifications for fluent operation onto the microcontroller, using the Arduino IDE environment.

The classification outcomes were subsequently accessible to the user through TCP/IP connection offered by the ESP8266 core of the radio module. Indeed, two separate hardware variants were used in the experiments, utilizing the ESP8266 chip. For the smart faucet (i.e., the water alert) case, a minimal ESP-01 board (Espressif Systems, Shanghai, China) was connected to the Arduino microcontroller, while for the water pump malfunction detection case, the NodeMCU board (Espressif Systems, Shanghai, China) was the preferable option. The ESP-01 comes with preinstalled firmware that offers modem-like communication commands. This fact makes its connection with the Arduino Nano 33 BLE Sense more complex, and thus the original code was substituted by a version that supports direct Wi-Fi and TCP/IP client/server functionality, via the ESP8266WiFi library. The lack of a USB port on the ESP-01 module requires an additional module for (re)programming it. For this reason, on the water pump system, the NodeMCU variant of the ESP8266 chip was used, which is more user-friendly. The NodeMCU board was primarily set up as a small web server hosting a straightforward HTML page with dynamic content reflecting the operational status of the water pump. Although more advanced networking methods were available, they were beyond the objectives of this study.

4.4. Monitoring Arrangements

To receive and inspect remote alerts through Wi-Fi, a basic monitoring application was created utilizing the MIT App Inventor environment [75]. This application utilized visual blocks and was designed to be run on an Android smartphone (or other Android device, e.g., tablet), which are commonly used by modern and especially young users [76]. Various algorithmic flavors utilizing TCP, UDP and HTTP messaging mechanisms were implemented and tested with the participation of the students [69]. Figure 5a depicts the smart phone interface design details using the MIT App Inventor environment, while Figure 5b presents indicative code blocks defining the smart phone application behavior. Initial experiments involved direct communication between the in situ sensors and the mobile device of the user. In this case, either the sensor node or the smart phone was acting as a Wi-Fi access point while the other device was set as a wireless client. At the next level, a dedicated access point was utilized, i.e., a TP-LINK TL-WR841N device (TP-LINK, Shenzhen, China). The last set of experiments involved a separate gateway/sink node, developed using a Raspberry Pi 3 Model B+, to tackle multiple sensors delivering misuse/malfunction alerts. This node intercepts data statements from the other sensor nodes and stores them into files, making them accessible through a TCP/IP-based service. This task is performed using Python and Linux shell scripts, techniques using IP sockets [77], and the activation of preexisting on the Raspberry Pi applications like the Apache web server [78]. Further networking optimization might require services and security/privacy settings that were beyond the scope of this work.

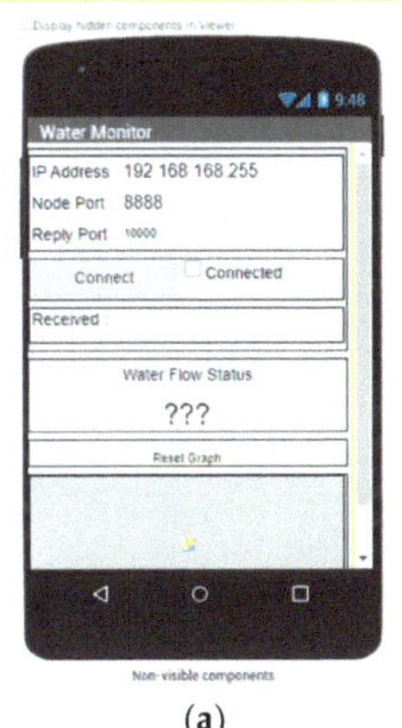

(a)

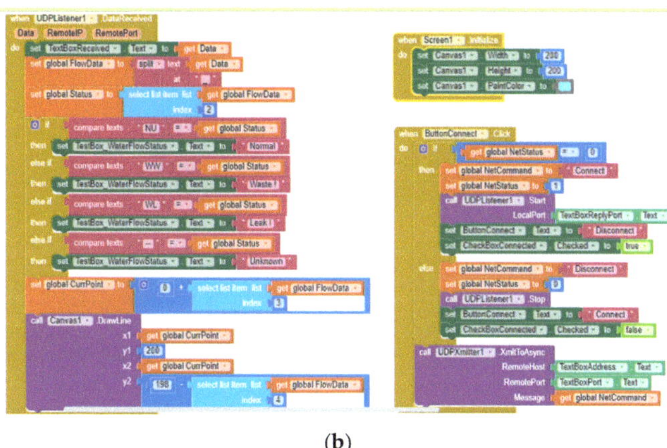

(b)

Figure 5. (**a**) Indicative smart phone interface design details using the MIT App Inventor environment; (**b**) Indicative code blocks defining the smart phone application behavior, using the MIT App Inventor Environment.

5. Results and Evaluation

The people getting involved into the corresponding evaluation study belonged to one or more of the following potential categories: students that participated in the construction, programming and training process of the two systems, students that were implementing small curricular projects of similar character, students that were explained how the pilot systems work, students that verified the functionality of the two systems by creating disturbance events (i.e., water waste or leak events or repositioning the valves of the pumping system) and by inspecting the corresponding results on their mobile phone screens. After finishing the above activities, the participants were asked to complete assessment forms and to provide potential remarks about the role of the proposed experimental systems incorporating machine learning techniques. In Section 5.1, technical details of these evaluation activities are highlighted, while Section 5.2 focuses on the educational counterpart.

5.1. Technical Aspect

Initially, the students assisted the process of data collection for both systems, that needed to be transferred to the edge impulse platform for the training of the machine learning models. Figure 6a,b depict the labeled raw data as shown from the environment of edge impulse during the features' generation process.

Following the training process, the confusion matrix of the model is generated automatically by the EI platform and the overall percentage accuracy is computed, as well as the testing accuracy of the model by utilizing data specifically reserved for this purpose is produced. Accuracy is the most used metric for evaluating classification models, often accompanied by a table in statistics known as the confusion matrix. Accuracy is the measure of the extent to which a model's predictions align with the actual reality, typically expressed as a percentage. In the realm of predictive analytics, the confusion matrix is a 2×2 table, providing information on the counts of true positives, false negatives, false positives, and true negatives [79–83].

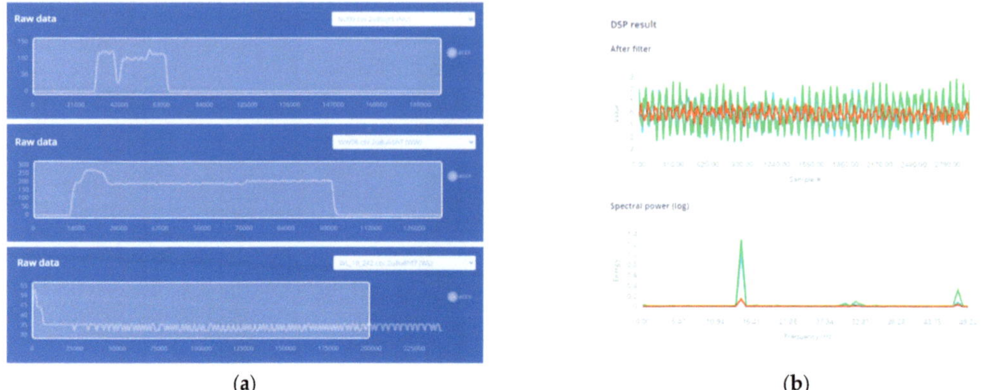

Figure 6. (**a**) Water flow sample data used for training the smart faucet system model; (**b**) Accelerometric sample data used for training the water-pump malfunction-detection system model.

In the case of the water-misuse neural-network model, according to the EI cloud environment, the NU category was correctly classified with an accuracy of 77.8%, the WW categories attained a 100% success rate and similarly the WL category reached 100% accuracy. These results led to a final model with an expected accuracy of 96.59% utilizing the testing data set as depicted in Figure 7a. In the next stage, the system was tested with actual episodes of water consumption (i.e., NU, WW, or WL) by appropriately rotating the tap head to allow the machine learning engine to classify the flow data gathered in segments of 200 consecutive values. The analysis of the collected data showed that the accuracy of the water-consumption prediction model was 91% when tested utilizing user-generated profiles with the recommended smart water metering system. It is noteworthy that the model was able to accurately identify unwanted WL profiles, with accuracy rates of up to 100%. However, some incorrect predictions were made, with the ML model confusing an actual WW situation as NU or WL.

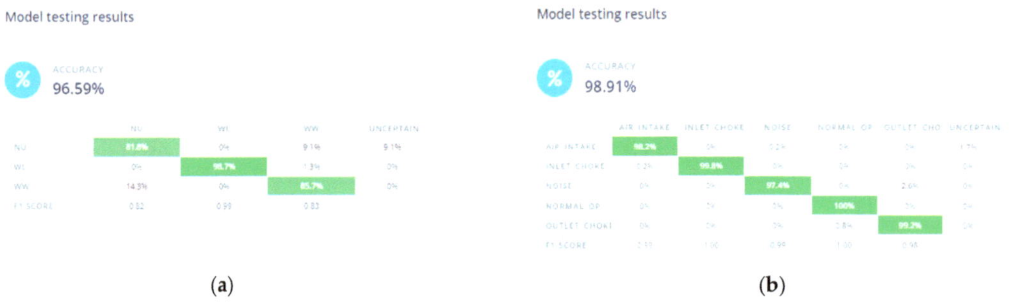

Figure 7. (**a**) Estimated performance of the machine-learning model for the smart faucet system; (**b**) Equivalent results for the water-pump malfunction-detection model.

Similarly, the accuracy of the system detecting water-pump malfunctions was evaluated by analyzing the platform results and conducting experiments with data that was not used in the training process. The overall accuracy percentage, as reported by the Edge Impulse, was 98.5%. The system achieved 100% accuracy for the normal operation category, while the accuracy rate for the air intake category was 97.2%, the accuracy rate for the inlet choke category was 99.8%, and for the outlet choke category, it was 96.7%, all according to the EI environment. These resulted to a total model accuracy score of 98.91% based on the testing dataset as shown in Figure 7b. Furthermore, a secondary assessment was conducted in order for the model to be tested on the actual system after classifying 1056 episodes. The analysis of the collected data showed that the overall accuracy reached 93.02% when the model was tested for malfunctions created by the students when manipulating the valves integrated into the system's design.

The inspection of the systems' performance was achieved through web interfaces and applications, as depicted in Figure 8a,b. The students used their smartphones in order to examine whether the model expected accuracy was reflected to the actual operation of each system. More specifically, students verified the functionality of the two systems by creating water misuse events (Figure 9a) via repositioning the valves of the pumping system (Figure 9b), and monitoring the corresponding results on their mobile phone screens. The recorded results indicated that 9 over 10 predictions were correct, which is in line with the performance of each machine-learning system, as explained in [29,30].

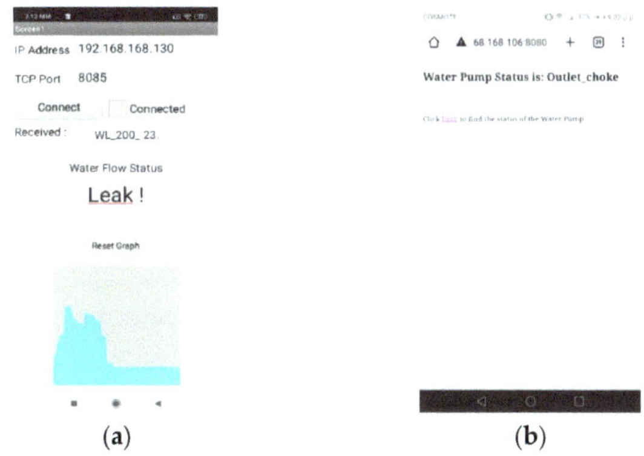

(a)   (b)

Figure 8. (**a**) Indicative smartphone screenshot during the students' inspection process, detecting a water leak event; (**b**) Equivalent smartphone screenshot reflecting the water-pump status during the in situ experiments.

(a)   (b)

Figure 9. (**a**) Students inspecting the operation of the prototype smart faucet system during the in-situ experiments; (**b**) Students were repositioning the valves of the smart pump system emulating malfunction events.

Finally, the students verified the actual performance also in terms of power efficiency using an accurate amperemeter connected in series with the electronic components of the sensor node. Each of the two detection systems consumed a few tenths of mA during its activity. According to experiments performed inside the university campus, the Wi-Fi radios were offering a range coverage of 100–150 m, at line-of-sight conditions. More specifically, the perceived signal strength of the ESP8266 radio was reported to the user and the connection link could sufficiently transfer data up to the −90 dBm border. Data rate of a few packets per second was enough to sustain the notification messages from the in-situ device toward the user. Although the focus of this study is more put on the machine learning functionality aspect of the proposed malfunction/misuse detection systems, these performance data is also necessary to be mentioned, as they are essential for understanding any IoT approach.

5.2. Educational Aspect

University students (both undergraduate and postgraduate) took part in all the stages of the development of the proposed systems, from the initial planning and design to the implementation, and final evaluation. The students engaged in those activities (55 participants in total), ranged from beginners (63.6% of them) to more experienced (36.4% of them), depending on their involvement in STEM activities, whether as part of their curriculum or extracurricular pursuits during the ten-month period of study. These participants were anonymously and consensually interviewed, through electronic forms [84] in Likert scale [85] questions to evaluate the entire process. An illustrative set of initial results, which are being collected and processed, is depicted in Figures 10–16. In the following charts, the bar height (vertical axis) illustrates the proportion of individuals with a particular level of agreement regarding the statement depicted above the chart. Blue bars refer to the water alert module and green bars pertain to the water pump fault detection system. The horizontal axis represents the characterization of opinion groups by a numerical scale ranging from 1 to 5, where the numbers 1 represent "Strongly Disagree", 2 stands for "Disagree", 3 denotes "Neutral", 4 indicates "Agree", and 5 signifies "Strongly Agree".

In all instances, the configuration was kept as open and inexpensive as possible to demonstrate high modularity and reusability of the units. This approach enables various educationally meaningful experimentation activities [63]. The survey results suggest that involvement in the entire activity was helpful for comprehending fundamental hardware and software topics, as well as for understanding the role of embedded systems (Figure 11a), and the proposed activities exhibit to be highly relevant to their courses at the university (Figure 10a). More specifically, most of the participants expressed that the activities acquaint them with machine learning and networking basics (Figures 10b and 11b, respectively), which are of great importance for giving intelligence to common systems.

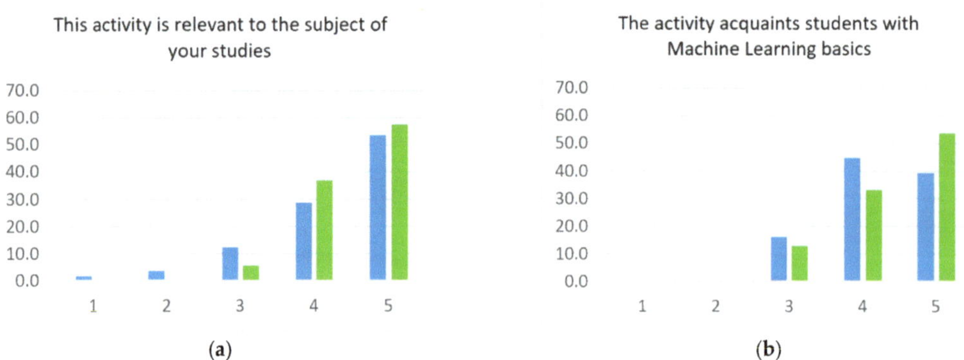

Figure 10. Participants' opinion about: (**a**) whether these activities are relevant to the subject of their studies; (**b**) the impact of the proposed activities on understanding machine-learning concepts.

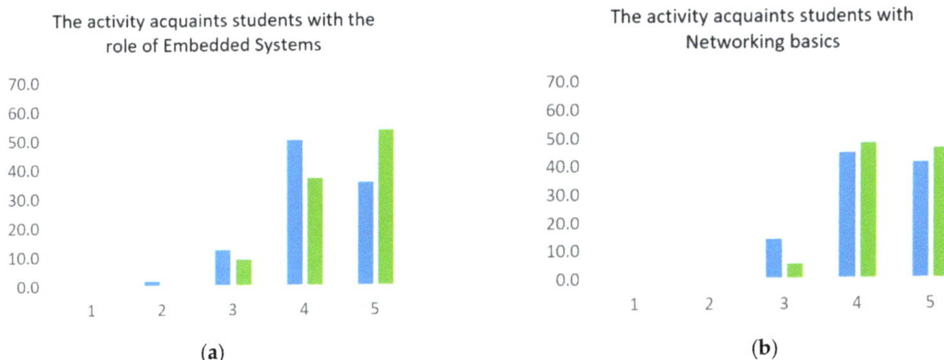

Figure 11. Participants' opinion about the impact of the presented activities to understanding: (**a**) the role of embedded systems; (**b**) networking basics.

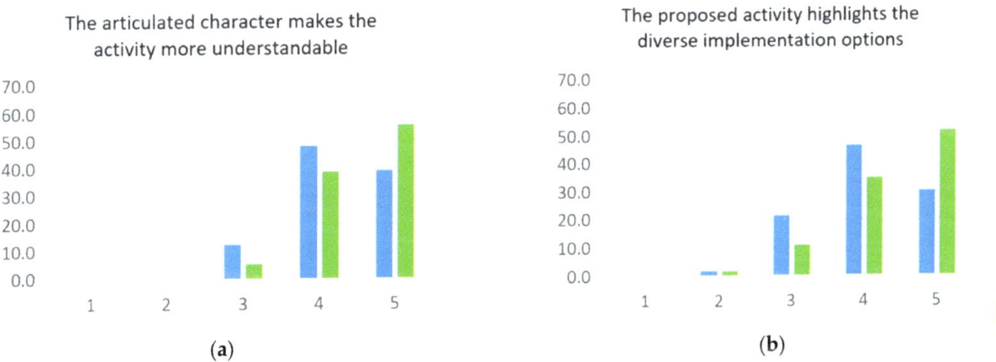

Figure 12. The views of the participants regarding the extent to which: (**a**) the articulated character makes the activities more understandable; (**b**) the proposed activities highlight the diverse implementation options.

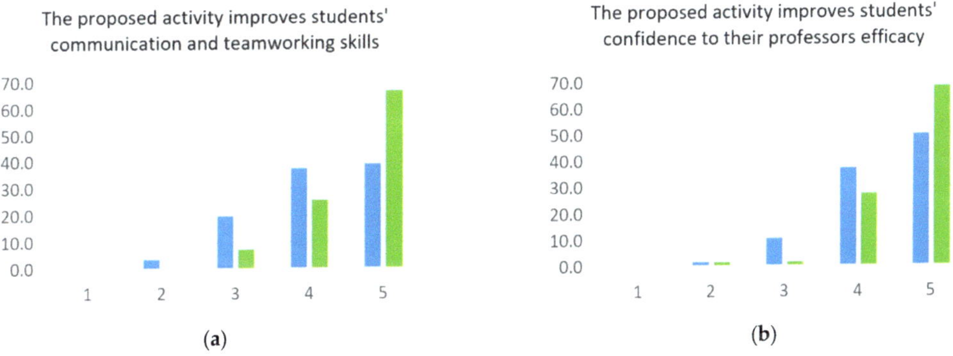

Figure 13. Participants' perspectives on the extent to which the presented activities improve: (**a**) students' communication and collaboration skills; (**b**) students' confidence in their professors' efficacy.

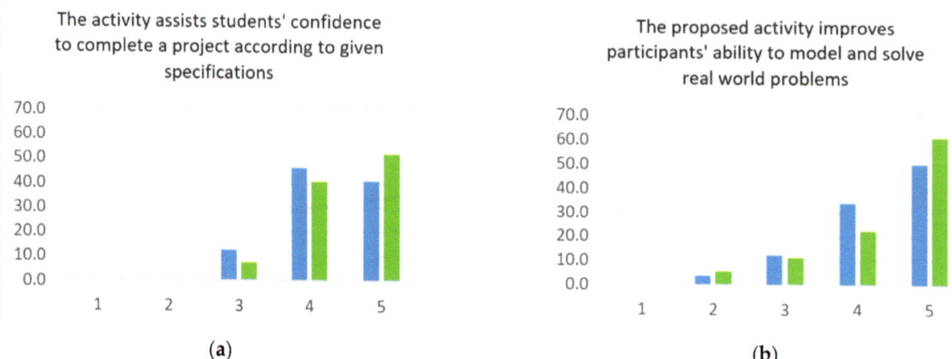

Figure 14. Viewpoint on the degree to which the presented activities: (**a**) Assist students' confidence to successfully finish a project based on provided specifications; (**b**) improve participants' ability to model and solve real-world problems.

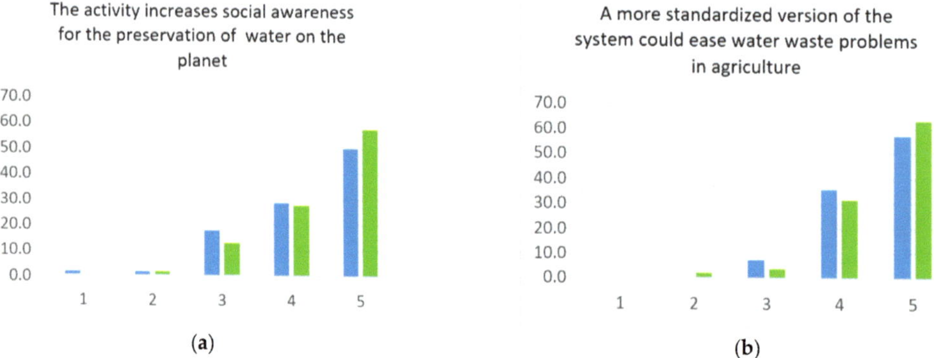

Figure 15. Viewpoint regarding the degree to which: (**a**) the proposed activities raise awareness in society about the importance of preserving water on the planet; (**b**) more standardized versions of the systems could ease water-waste problems in agriculture.

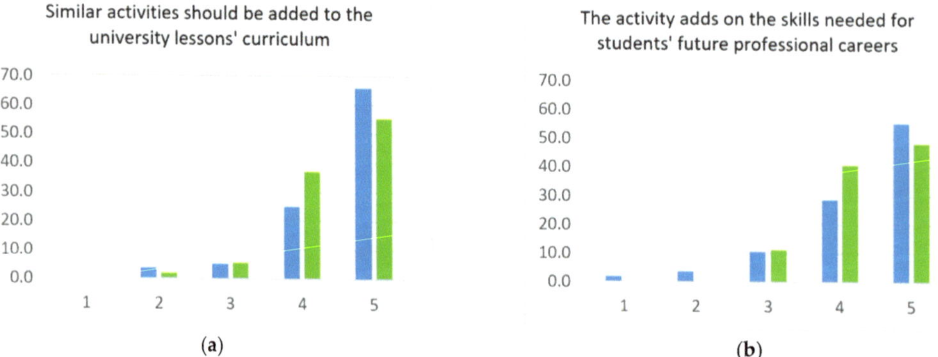

Figure 16. Opinion about: (**a**) whether comparable activities should be incorporated into the university curriculum; (**b**) the influence of the suggested activities on students' acquisition of skills for their professional careers.

Moreover, the respondents found that the articulated character of the systems was more understandable (Figure 12a), while at the same time, the diverse implementation options were emphasized (Figure 12b). Similarly, participants' viewpoints regarding the contribution of the presented activity to their soft skills acquisition were very positive. The involvement in the proposed activities improved students' communication and team working skills (Figure 13a) and enhanced students' confidence in their educators' efficacy (Figure 13b).

It is noteworthy that the detailed hardware and software configurations of the project served as lens, highlighting the challenges in the ongoing upgrade process. Based on the feedback from the respondents, the overall activities boosted their confidence in completing a task according to given specifications (Figure 14a), while improving their ability to model and solve real-world problems (Figure 14b). Significantly for these benefits, besides the Arduino programming community and the user-friendly Edge Impulse platform, was the facilitation of the MIT App Inventor cloud-based environment, enabling code sharing and rapid prototyping, along with the raspberry pi, which supports various programming sessions and allows for examining the behavior of the embedded system.

As explained, in Figure 15a, participants believe that this initiative enhances social awareness regarding the conservation of water worldwide, and a better standardized variant of the system could alleviate water wastage issues in agriculture. (Figure 15b), which are of great importance, as explained in the previous sections. Finally, it is worth highlighting that most participants believed that similar topics should be included in the university courses curricula (Figure 16a), while they also believe that the proposed activities triggered students' interests and were pertinent to the skills required for their future professional careers. (Figure 16b).

The suggested approach was adjusted to introduce and foster communication among students about the fundamentals of machine learning, which is an essential process for many agricultural operations and are connected with other contemporary technological methods of the digital era, such as networking. It must be noted that, based on the literature review and the current state of agricultural management, it can be inferred that there are limited specialized educational courses addressing these concerns [41]. The findings presented align with and build upon previous educational discoveries in the field [63,66,86], as they fulfill the objectives of agri-food professionals involved in the shift towards sustainable agriculture. Indeed, according to the study presented in [87], upcoming professionals require skills that foster an attitude based on diversity and the inclusion of various information, methods, and experiences, along with the ability to respond and proactively engage in a dynamic world.

5.3. Further Discussion

Throughout the activities presented, the instructors maintained a supportive and unobtrusive role, offering guidance and advice when requested. During the laboratory lessons, professors aimed to capture students' interests by gradually introducing topics in terms of complexity and rewarding students' daily improvement. The fact that the more experienced students, serving as mentors, was also of paramount significance. The optimistic attitude towards the different technical challenges in the process of upgrading water equipment proved to be the most effective paradigm, fostering further creativity and improvement.

Some non-machine-learning critical parts should be programmed and optimized independently and the entire program (sketch) for the Arduino Nano 33 BLE Sense unit should be kept as straightforward as possible. This is because compiling sketches for the Arduino Nano 33 BLE Sense unit requires extra time than those intended for the standard Arduino Uno platform. Furthermore, considering the implementation of the artificial neural network (ANN) model generated by Edge Impulse, the compilation duration variability was extended, i.e., to sometimes exceed a 15-min period. For these reasons, the decision being taken to utilize a second, inexpensive, and faster-to-program Arduino-based device (such as the NodeMCU board), to accomplish the auxiliary tasks with minimal interaction with the Arduino Nano 33 BLE Sense unit, facilitated the experiments.

Furthermore, favoring a slight overfitting of the neural-network models aimed the students to understand the connection between the data used for training and the final behavior. Additionally, the simplified models aided the students in comprehending the entire training process and the individual stages being necessary. These settings allowed them to experiment with diverse features and generate more than one machine-learning model variants to select the most suitable one, in reasonable time and processing cost from an educational approach perspective. In general, as the possibility of experiencing failures in either the educational or the technical settings of the activities being discussed is always present, the overall approach should be kept as simple as possible.

Moreover, a challenge in this rapidly evolving area is the capability of using and being familiar with new technologies. For this reason, potential educators getting involved should be able to proceed beyond the narrow limits of their specialty, in order to organize and assist student teams, and to provide advice for solving the difficulties that arise. Added to this, they should be able to keep meticulous records and documentation of the overall process, a practice that fosters the reproducibility of the good practices being experienced. The latter attitude is also helpful for not losing valuable educational and technical resources typically acquired from the cloud, which are frequently liable to drastic changes due to their innovative character.

The selection of components was not as optimal as possible from a commercial production perspective, i.e., trying to find a good compromise between cost minimization, decent system performance, and educational friendliness. The latter (and most important) objective was favoring well-documented components with high modularity and reusability potential and comparatively easy assembling. In this regard, the utilization of the Arduino Nano 33 BLE Sense, the Edge Impulse platform, the MIT App Inventor and the Arduino IDE environment was fully justified.

In this regard, incorporating Wi-Fi technologies and smart phones was a good practice from the educational point of view, as young people tend to adore these devices and are familiar with the corresponding wireless network settings. Nevertheless, for the future, realistic IoT experiences would require the engagement of more optimized technologies, such as LoRaWAN radios (LoRa) [88], and the implementation of sleeping/waking-up functionality on the sensor nodes, along with a more fluent monitoring software, providing access beyond the limits of the university laboratory wireless local area network (WLAN).

It is worth mentioning that extremely useful feedback, referring to the systems being investigated, was provided by the students that are traditionally being more capable for unbiased thoughts, compared to their professors. More specifically, some participants proposed the system to incorporate functions that stop the pump from working, through a relay, on malfunction detection, or to close the water supply to the faucet via an electric valve, on water waste or leak events. The subject of efficiently powering the smart systems being installed in situ, was also a fruitful field for inquiries, as students proposed solutions utilizing the water flow itself (via a micro turbine) for generating the necessary current for the smart faucet system, or solar panels solution for both systems and even for their actuating part (i.e., for the pump, the relays, and the valves).

By harnessing the capabilities of machine learning and artificial intelligence, we can not only monitor water preservation more effectively but also significantly improve the availability of pertinent information. These advanced technologies will enable us to analyze vast datasets in real-time, identify patterns, and make predictions related to water usage and conservation practices. This enhanced level of data-driven insights can empower decision makers, researchers, and environmentalists to devise more informed strategies for sustainable water management. Ultimately, the integration of machine learning and AI systems in monitoring water preservation not only increases efficiency but also lays the foundation for smarter, more adaptive water-resource management in the face of evolving environmental challenges.

The directions for implementing an affordable water usage alert system and water pump fault detection system, despite being in their early stages, can be beneficial for various real-world scenarios in both urban and rural areas, which is an encouraging outcome of the research conducted. To expand the scope of the research, the proposed systems and methodology will be further optimized and evaluated from both technical and educational perspectives. This will provide a more comprehensive set of results and practical solutions for various real-world applications. In this direction, the development of a commercial standards version for the discussed water alert module is being considered and pump's fault detection system will be a significant priority. Plans involve additional system enhancements, utilizing similar cost-effective components, and/or implementing other structures following a similar upgrading approach. The motivation is to enable the students of today and professionals of the future to experiment with a plethora of real-world challenges that they will face in their careers and contribute to a more sustainable future.

6. Conclusions

In this paper, two systems are demonstrated aiming to orchestrate educationally meaningful activities, for higher education, focused on water preservation and sustainability. In greater detail, a retired water pump for agricultural premises and a faucet were utilized and transformed into smart IoT systems, with the assistance of machine learning and embedded low-cost microcontrollers with networking capabilities, graphical user interfaces and smartphone devices. The greatest challenge was to form the necessary paradigm without sacrificing the real-world application suitability potential that these systems have and simultaneously to keep implementation reproducibility and cost at reasonable levels. Widely available hardware and software components were selected exhibiting easy-to-use character and fluent documentation.

According to the initial survey findings, the case being presented suggest that the approach is effective in achieving the increase of social awareness about water conservation while enhancing students' understanding of IoT and ML matters that are crucial for their future careers. The proposed approach assisted the participants in the educational activities to acquire multidisciplinary benefits, i.e., gaining more technical knowledge while simultaneously implementing applications that contribute to sustainability objectives. University students of little technological background were assisted to demystify cutting-edge technologies, and to bridge the gap between small-sized educational constructions and real-size systems. According to the survey findings, similar activities should be incorporated into the curricula of educational institutions and foster the future professional careers of the participants.

Author Contributions: Conceptualization, M.K. and D.L.; methodology, D.L.; investigation, M.K. and D.L.; software, D.L.; prototyping, D.L.; validation, M.K., D.L. and K.-A.L.; data curation, M.K., D.L. and K.-A.L.; writing—original draft preparation, M.K. and D.L.; writing—review and editing, M.K., D.L. and E.S.; visualization, M.K. and D.L.; supervision, K.G.A. and C.M. All authors have read and agreed to the published version of the manuscript.

Funding: This research received no external funding.

Institutional Review Board Statement: Ethical review and approval were waived for this study, due to not involving identifiable personal nor sensitive data.

Informed Consent Statement: Students' consent was waived due to not involving identifiable personal nor sensitive data.

Data Availability Statement: The data presented in this study are available upon request from the corresponding author.

Acknowledgments: The authors would like to thank the faculty and the students of the Department of Natural Resources Management and Agricultural Engineering of the Agricultural University of Athens, Greece, for their assistance during the implementation and/or evaluation activities being presented.

Conflicts of Interest: The authors declare no conflicts of interest.

References

1. Tan, J.T.C.; Iocchi, L.; Eguchi, A.; Okada, H. Bridging Robotics Education between High School and University: RoboCup@Home Education. In Proceedings of the 2019 IEEE AFRICON, Accra, Ghana, 25–27 September 2019. [CrossRef]
2. Papadakis, S.; Kalogiannakis, M. (Eds.) *Handbook of Research on Using Educational Robotics to Facilitate Student Learning. Advances in Educational Technologies and Instructional Design*; IGI Global: Hershey, PA, USA, 2021. [CrossRef]
3. Markham, T. Project Based Learning. *Teach. Libr.* **2011**, *39*, 38–42.
4. Anwar, S.; Bascou, N.A.; Menekse, M.; Kardgar, A. A Systematic Review of Studies on Educational Robotics. *J. Pre-Coll. Eng. Educ. Res.* **2019**, *9*, 2. [CrossRef]
5. Scaradozzi, D.; Cesaretti, L.; Screpanti, L.; Mangina, E. Identification of the Students Learning Process During Education Robotics Activities. *Front. Robot. AI* **2020**, *7*, 21. [CrossRef] [PubMed]
6. Doran, M.V.; Clark, G.W. Enhancing Robotic Experiences throughout the Computing Curriculum. In Proceedings of the SIGCSE'18: 49th ACM Technical Symposium on Computer Science Education, Baltimore, MD, USA, 21–24 February 2018; pp. 368–371.
7. Sapounidis, T.; Alimisis, D. Educational Robotics Curricula: Current Trends and Shortcomings. In *Education in & with Robotics to Foster 21st-Century Skills*; Springer: Cham, Switzerland, 2021; pp. 127–138. [CrossRef]
8. Sahin, Y.; Celikkan, U. Information Technology Asymmetry and Gaps between Higher Education Institutions and Industry. *J. Inf. Technol. Educ. Res.* **2020**, *19*, 339–365. [CrossRef]
9. Conde, J.; López-Pernas, S.; Pozo, A.; Munoz-Arcentales, A.; Huecas, G.; Alonso, Á. Bridging the Gap between Academia and Industry through Students' Contributions to the FIWARE European Open-Source Initiative: A Pilot Study. *Electronics* **2021**, *10*, 1523. [CrossRef]
10. Nascimento, D.; Chavez, C.; Bittencourt, R. The Adoption of Open Source Projects in Engineering Education: A Real Software Development Experience. In Proceedings of the 48th IEEE Frontiers in Education Conference (FIE), San Jose, CA, USA, 3–6 October 2018; pp. 1–9.
11. Chung, C.A. A Cost-Effective Approach for the Development of an Integrated PC-PLC-Robot System for Industrial Engineering Education. *IEEE Trans. Educ.* **1998**, *41*, 306–310. [CrossRef]
12. Vargas, H.; Sanchez Moreno, J.; Jara, C.A.; Candelas, F.A.; Torres, F.; Dormido, S. A Network of Automatic Control Web-Based Laboratories. *IEEE Trans. Learn. Technol.* **2011**, *4*, 197–208. [CrossRef]
13. Phan, M.-H.; Ngo, H.Q.T. A Multidisciplinary Mechatronics Program: From Project-Based Learning to a Community-Based Approach on an Open Platform. *Electronics* **2020**, *9*, 954. [CrossRef]
14. Jensen, K.; Larsen, M.; Nielsen, S.; Larsen, L.; Olsen, K.; Jørgensen, R. Towards an Open Software Platform for Field Robots in Precision Agriculture. *Robotics* **2014**, *3*, 207–234. [CrossRef]
15. Alimisis, D.; Alimisi, R.; Loukatos, D.; Zoulias, E. Introducing Maker Movement in Educational Robotics: Beyond Prefabricated Robots and "Black Boxes". In *Smart Learning with Educational Robotics*; Daniela, L., Ed.; Springer: Cham, Switzerland, 2019. [CrossRef]
16. Benavides, L.M.C.; Tamayo Arias, J.A.; Arango Serna, M.D.; Branch Bedoya, J.W.; Burgos, D. Digital Transformation in Higher Education Institutions: A Systematic Literature Review. *Sensors* **2020**, *20*, 3291. [CrossRef]
17. Bianchi, G.; Pisiotis, U.; Cabrera Giraldez, M. *GreenComp*; Publications Office of the European Union: Luxembourg, 2022.
18. United Nations Brundtland Commission. 1987. Available online: https://www.un.org/en/academic-impact/sustainability (accessed on 20 October 2023).
19. Ramakrishna, S.; Jose, R. Addressing sustainability gaps. *Sci. Total Environ.* **2022**, *806*, 151208. [CrossRef] [PubMed]
20. Roomi, M.A.; Saiz-Alvarez, J.M.; Coduras, A. Measuring sustainable entrepreneurship and eco-innovation: A methodological proposal for the Global Entrepreneurship Monitor (GEM). *Sustainability* **2021**, *13*, 4056. [CrossRef]
21. Prabakaran, M. Historical appropriation of epistemological values: A goal ahead for higher education. *High. Educ. Future* **2020**, *7*, 67–81. [CrossRef]
22. Abduganiev, O.I.; Abdurakhmanov, G.Z. Ecological education for the purposes of sustainable development. *Am. J. Soc. Sci. Educ. Innov.* **2020**, *2*, 280–284. [CrossRef]
23. Jeronen, E. Sustainable Education. In *Encyclopedia of Sustainable Management*; Idowu, S., Schmidpeter, R., Capaldi, N., Zu, L., Del Baldo, M., Abreu, R., Eds.; Springer: Cham, Switzerland, 2022. [CrossRef]
24. Wade, R. Education for sustainability: Challenges and opportunities. In *Policy & Practice: A Development Education Review*; Coriddi, J., Ed.; NI Ltd.: Lisburn, Northern Ireland, 2008; pp. 30–48.
25. FAO. *The State of the World's Land and Water Resources for Food and Agriculture: Managing Systems at Risk*; FAO: Rome, Italy, 2018.
26. Paraforos, D.S.; Griepentrog, H.W. Digital farming and field robotics: Internet of things, cloud computing, and big data. In *Fundamentals of Agricultural and Field Robotics*; Springer: Cham, Switzerland, 2021; pp. 365–385.
27. Liakos, K.G.; Busato, P.; Moshou, D.; Pearson, S.; Bochtis, D. Machine Learning in Agriculture: A Review. *Sensors* **2018**, *18*, 2674. [CrossRef] [PubMed]
28. Shi, W.; Cao, J.; Zhang, Q.; Li, Y.; Xu, L. Edge computing: Vision and challenges. *IEEE Internet Things J.* **2016**, *3*, 637–646. [CrossRef]
29. Loukatos, D.; Lygkoura, K.-A.; Maraveas, C.; Arvanitis, K.G. Enriching IoT Modules with Edge AI Functionality to Detect Water Misuse Events in a Decentralized Manner. *Sensors* **2022**, *22*, 4874. [CrossRef] [PubMed]

30. Loukatos, D.; Kondoyanni, M.; Alexopoulos, G.; Maraveas, C.; Arvanitis, K.G. On-Device Intelligence for Malfunction Detection of Water Pump Equipment in Agricultural Premises: Feasibility and Experimentation. *Sensors* **2023**, *23*, 839. [CrossRef] [PubMed]
31. Soma, T.; Nuckchady, B. Communicating the Benefits and Risks of Digital Agriculture Technologies: Perspectives on the Future of Digital Agricultural Education and Training. *Front. Commun.* **2021**, *6*, 762201. [CrossRef]
32. Trisha, V.; Golick, D.; Stains, M. Educational Technologies and Instructional Practices in Agricultural Sciences: Leveraging the Technological Pedagogical Content Knowledge (TPACK) Framework to Critically Review the Literature. *NACTA J.* **2018**, *62*, 65–76. Available online: https://www.jstor.org/stable/90021575 (accessed on 10 January 2023).
33. Migliorini, P.; Wezel, A.; Veromann, E.; Strassner, C.; Średnicka-Tober, D.; Kahl, J.; Bügel, S.; Briz, T.; Kazimierczak, R.; Brives, H.; et al. Students' Knowledge and Expectations about Sustainable Food Systems in Higher Education. *Int. J. Sustain. High. Educ.* **2020**, *21*, 1087–1110. [CrossRef]
34. Education for Sustainable Development (ESD). 2022. Available online: https://en.unesco.org/themes/education/sdgs/material (accessed on 3 July 2023).
35. Castellanos, P.M.A.; Queiruga-Dios, A. From environmental education to education for sustainable development in higher education: A systematic review. *Int. J. Sustain. High. Educ.* **2021**, *23*, 622–644. [CrossRef]
36. Lozano, R.; Lozano, F.J.; Mulder, K.; Huisingh, D.; Waas, T. Advancing higher education for sustainable development: International insights and critical reflections. *J. Clean. Prod.* **2013**, *48*, 3–9. [CrossRef]
37. Luna-Krauletz, M.D.; Juárez-Hernández, L.G.; Clark-Tapia, R.; Súcar-Súccar, S.T.; Alfonso-Corrado, C. Environmental Education for Sustainability in Higher Education Institutions: Design of an Instrument for Its Evaluation. *Sustainability* **2021**, *13*, 7129. [CrossRef]
38. Rodríguez-Abitia, G.; Martínez-Pérez, S.; Ramirez-Montoya, M.S.; Lopez-Caudana, E. Digital Gap in Universities and Challenges for Quality Education: A Diagnostic Study in Mexico and Spain. *Sustainability* **2020**, *12*, 9069. [CrossRef]
39. Loukatos, D.; Androulidakis, N.; Arvanitis, K.G.; Peppas, K.P.; Chondrogiannis, E. Using Open Tools to Transform Retired Equipment into Powerful Engineering Education Instruments: A Smart Agri-IoT Control Example. *Electronics* **2022**, *11*, 855. [CrossRef]
40. Chatzopoulos, A.; Tzerachoglou, A.; Priniotakis, G.; Papoutsidakis, M.; Drosos, C.; Symeonaki, E. Using STEM to Educate Engineers about Sustainability: A Case Study in Mechatronics Teaching and Building a Mobile Robot Using Upcycled and Recycled Materials. *Sustainability* **2023**, *15*, 15187. [CrossRef]
41. Mijailović, Đ.; Đorđević, A.; Stefanovic, M.; Vidojević, D.; Gazizulina, A.; Projović, D. A Cloud-Based with Microcontroller Platforms System Designed to Educate Students within Digitalization and the Industry 4.0 Paradigm. *Sustainability* **2021**, *13*, 12396. [CrossRef]
42. Laupichler, M.C.; Aster, A.; Perschewski, J.-O.; Schleiss, J. Evaluating AI Courses: A Valid and Reliable Instrument for Assessing Artificial-Intelligence Learning through Comparative Self-Assessment. *Educ. Sci.* **2023**, *13*, 978. [CrossRef]
43. Su, J.; Yang, W. Artificial intelligence in early childhood education: A scoping review. *Comput. Educ. Artif. Intell.* **2022**, *3*, 100049. [CrossRef]
44. Casal-Otero, L.; Catala, A.; Fernández-Morante, C.; Taboada, M.; Cebreiro, B.; Barro, S. AI literacy in K-12: A systematic literature review. *Int. J. STEM Educ.* **2023**, *10*, 29. [CrossRef]
45. Ng, D.T.K.; Leung, J.K.L.; Su, M.J.; Yim, I.H.Y.; Qiao, M.S.; Chu, S.K.W. *AI Literacy in K-16 Classrooms*; Springer International Publishing: Berlin/Heidelberg, Germany, 2023.
46. Southworth, J.; Migliaccio, K.; Glover, J.; Reed, D.; McCarty, C.; Brendemuhl, J.; Thomas, A. Developing a model for AI Across the curriculum: Transforming the higher education landscape via innovation in AI literacy. *Comput. Educ. Artif. Intell.* **2023**, *4*, 100127. [CrossRef]
47. Laupichler, M.C.; Aster, A.; Schirch, J.; Raupach, T. Artificial intelligence literacy in higher and adult education: A scoping literature review. *Comput. Educ. Artif. Intell.* **2022**, *3*, 100101. [CrossRef]
48. Garcia Lopez, P.; Montresor, A.; Epema, D.; Datta, A.; Higashino, T.; Iamnitchi, A.; Barcellos, M.; Felber, P.; Riviere, E. Edge-centric computing: Vision and challenges. *ACM SIGCOMM Comput. Commun. Rev.* **2015**, *45*, 37–42. [CrossRef]
49. Dineva, K.; Atanasova, T. Systematic Look at Machine Learning Algorithms–Advantages, Disadvantages and Practical Applications. *Int. Multidiscip. Sci. GeoConference SGEM* **2020**, *20*, 317–324.
50. European Commission; Directorate-General for Communications Networks, Content and Technology; Blind, K.; Muto, S.; Pätsch, S.; Schubert, T. *The Impact of Open Source Software and Hardware on Technological Independence, Competitiveness and Innovation in the EU Economy: Final Study Report*; Publications Office: Luxembourg, 2021. Available online: https://data.europa.eu/doi/10.2759/430161 (accessed on 15 December 2023).
51. Sustainable Development Goals. Available online: https://sdgs.un.org/goals (accessed on 6 December 2021).
52. Wenger, E.; McDermott, R.A.; Snyder, W. *Cultivating Communities of Practice: A Guide to Managing Knowledge*; Harvard Business School Press: Boston, MA, USA, 2002.
53. Corlu, M.; Capraro, R.M.; Capraro, M.M. Introducing STEM Education: Implications for Educating Our Teachers for the Age of Innovation. *Educ. Sci.* **2014**, *39*, 74–85.
54. Hallström, J.; Schönborn, K.J. Models and modelling for authentic STEM education: Reinforcing the argument. *Int. J. STEM Educ.* **2019**, *6*, 22. [CrossRef]

55. Fisher-Maltese, C.; Zimmerman, T.D. A garden-based approach to teaching life science produces shifts in students' attitudes toward the environment. *Int. J. Environ. Sci. Educ.* **2015**, *10*, 51–66.
56. Stubbs, E.A.; Myers, B.E. Multiple Case Study of STEM in School-based Agricultural Education. *J. Agric. Educ.* **2015**, *56*, 188–203. [CrossRef]
57. Borges, J.; Dias, T.G.; Cunha, J.F. A new group-formation method for student projects. *Eur. J. Eng. Educ.* **2019**, *34*, 573–585. [CrossRef]
58. Portuguez Castro, M.; Gómez Zermeño, M.G. Challenge Based Learning: Innovative Pedagogy for Sustainability through e-Learning in Higher Education. *Sustainability* **2020**, *12*, 4063. [CrossRef]
59. Gutiérrez-Martínez, Y.; Bustamante-Bello, R.; Navarro-Tuch, S.A.; López-Aguilar, A.A.; Molina, A.; Álvarez-Icaza Longoria, I. A Challenge-Based Learning Experience in Industrial Engineering in the Framework of Education 4.0. *Sustainability* **2021**, *13*, 9867. [CrossRef]
60. van den Beemt, A.; Vázquez-Villegas, P.; Gómez Puente, S.; O'Riordan, F.; Gormley, C.; Chiang, F.-K.; Leng, C.; Caratozzolo, P.; Zavala, G.; Membrillo-Hernández, J. Taking the Challenge: An Exploratory Study of the Challenge-Based Learning Context in Higher Education Institutions across Three Different Continents. *Educ. Sci.* **2023**, *13*, 234. [CrossRef]
61. Gudonienė, D.; Paulauskaitė-Tarasevičienė, A.; Daunorienė, A.; Sukackė, V. A Case Study on Emerging Learning Pathways in SDG-Focused Engineering Studies through Applying CBL. *Sustainability* **2021**, *13*, 8495. [CrossRef]
62. Smith, B.L.; MacGregor, J.T. What is collaborative learning. In *Collaborative Learning: A Sourcebook for Higher Education*; Goodsell, A.S., Maher, M.R., Tinto, V., Eds.; National Center on Postsecondary Teaching, Learning, & Assessment, Syracuse University: Syracuse, NY, USA, 1992.
63. Loukatos, D.; Kondoyanni, M.; Kyrtopoulos, I.-V.; Arvanitis, K.G. Enhanced Robots as Tools for Assisting Agricultural Engineering Students' Development. *Electronics* **2022**, *11*, 755. [CrossRef]
64. Sukackė, V.; Guerra, A.O.P.d.C.; Ellinger, D.; Carlos, V.; Petronienė, S.; Gaižiūnienė, L.; Blanch, S.; Marbà-Tallada, A.; Brose, A. Towards Active Evidence-Based Learning in Engineering Education: A Systematic Literature Review of PBL, PjBL, and CBL. *Sustainability* **2022**, *14*, 13955. [CrossRef]
65. King, A. Structuring Peer Interaction to Promote High-Level Cognitive Processing. *Theory Pract.* **2002**, *41*, 33–39. [CrossRef]
66. He, Y.; Liang, L. Application of Robotics in Higher Education in Industry 4.0 Era. *Univers. J. Educ. Res.* **2019**, *7*, 1612–1622. [CrossRef]
67. Arduino Nano 33 BLE Sense. Overview of the Arduino Nano 33 BLE Sense Microcontroller Unit. 2022. Available online: https://store.arduino.cc/products/arduino-nano-33-ble-sense (accessed on 25 February 2023).
68. Arduino Uno. Arduino Uno Board Description on the Official Arduino Site. 2021. Available online: https://store.arduino.cc/arduino-uno-rev3 (accessed on 20 February 2023).
69. ESP8266. The ESP8266 Low-Cost Wi-Fi Microchip. 2023. Available online: https://en.wikipedia.org/wiki/ESP8266 (accessed on 20 February 2023).
70. Arduino Software IDE. 2022. Available online: https://www.arduino.cc/en/Guide/Environment (accessed on 20 February 2023).
71. Burns, E.; Bruke, J. What is an Artificial Neural Network (ANN)? SearchEnterpriseAI. 2021. Available online: https://searchenterpriseai.techtarget.com/definition/neural-network (accessed on 1 December 2023).
72. EdgeImpulse. 2021. Available online: https://www.edgeimpulse.com/ (accessed on 2 December 2021).
73. Twinschip. Water Flow Meter. 2023. Available online: https://www.twinschip.com/Water-Flow%20Sensor-Control-Effect-Flowmeter-Hall--YF-S201 (accessed on 20 February 2023).
74. ESP8266 UDP Send & Receive. 2020. Available online: https://siytek.com/esp8266-udp-send-receive/ (accessed on 20 February 2023).
75. MIT App Inventor Programming Environment. Available online: https://appinventor.mit.edu/ (accessed on 20 February 2023).
76. Krouska, A.; Troussas, C.; Sgouropoulou, C. Mobile game-based learning as a solution in COVID-19 era: Modeling the pedagogical affordance and student interactions. *Educ. Inf. Technol.* **2021**, *27*, 229–241. [CrossRef]
77. UDP—Client and Server Example Programs in Python. 2023. Available online: https://pythontic.com/modules/socket/udp-client-server-example (accessed on 20 February 2023).
78. Raspberry Pi Apache Web Server Setup. 2022. Available online: https://pimylifeup.com/raspberry-pi-apache/ (accessed on 20 February 2023).
79. Fawcett, T. An introduction to ROC analysis. *Pattern Recogn. Lett.* **2006**, *27*, 861–874. [CrossRef]
80. Tharwat, A. Classification assessment methods. *Appl. Comput. Inform.* **2018**, *17*, 168–192. [CrossRef]
81. Powers, D.M.W. Evaluation: From Precision, Recall and F-Measure to ROC, Informedness, Markedness & Correlation. *J. Mach. Learn. Technol.* **2011**, *2*, 37–63.
82. Shultz, T.R.; Fahlman, S.E.; Craw, S.; Andritsos, P.; Tsaparas, P.; Silva, R.; Wiegand, R.P. Confusion matrix. *Encycl. Mach. Learn.* **2011**, *61*, 209.
83. Kakas, A.C.; Cohn, D.; Dasgupta, S.; Barto, A.G.; Carpenter, G.A.; Grossberg, S.; Webb, G.I. Accuracy. In *Encyclopedia of Machine Learning*; Springer: Boston, MA, USA, 2011; pp. 9–10.
84. Google Forms. Repository of Guidance and Tools for the Google Forms. 2021. Available online: https://www.google.com/forms/about/ (accessed on 28 September 2021).
85. Likert, R. A Technique for the Measurement of Attitudes. *Arch. Psychol.* **1932**, *140*, 55.

86. Cardín Pedrosa, M.; Marey-Perez, M.; Cuesta, T.; Álvarez, C.J. Agricultural Engineering Education in Spain. *Int. J. Eng. Educ.* **2014**, *30*, 1023–1035.
87. Sørensen, L.B.; Germundsson, L.B.; Hansen, S.R.; Rojas, C.; Kristensen, N.H. What Skills Do Agricultural Professionals Need in the Transition towards a Sustainable Agriculture? A Qualitative Literature Review. *Sustainability* **2021**, *13*, 13556. [CrossRef]
88. LoRa32u4. The LoRa32u4 Module Description. 2022. Available online: https://www.diymalls.com/LoRa32u4-II-Lora-Development-Board-868mhz-915mhz-Lora-Module (accessed on 25 July 2023).

 *sustainability*

Article

Implementation of Integrated Environmental Management and Its Specialized Engineering Education in Korea: A Case Study

Da-Som Park [1] , Moon-Seok Kang [1], Chan-Byeong Chae [1], Young Sunwoo [2] and Ki-Ho Hong [2,*]

[1]   Department of Environmental Engineering, Konkuk University, 120 Neungdong-ro, Seoul 05029, Republic of Korea; shn02022@gmail.com (D.-S.P.); msgang01@gmail.com (M.-S.K.); lopsup@naver.com (C.-B.C.)

[2]   Department of Civil and Environmental Engineering, Konkuk University, 120 Neungdong-ro, Seoul 05029, Republic of Korea; ysunwoo@konkuk.ac.kr

*   Correspondence: khhong@konkuk.ac.kr; Tel.: +82-2-450-4047

Abstract: Integrated environmental management (IEM) is an effective approach that comprehensively reviews the impact of pollutants emitted from a facility on the surrounding environment and minimizes pollution emissions through optimal, economically feasible means. The IEM system in Korea, as derived from countries in the European Union and other advanced countries, has been in force since 2017. This study presents the primary components and features of the IEM system in Korea. IEM specialized education is aimed at introducing expertise in the field and equipping learners with various skills upon completion, including in-depth research skills, to meet Korea's stringent requirements. Regular performance checks and surveys of program participants are conducted to confirm objective results. Based on these data, the program is improved according to expert opinions and contributes numerous engineering inputs for environmentally sustainable management. The education system can be used to develop and apply processes for training system operation personnel in many countries interested in adapting IEM, as well as in Europe, which already utilizes an IEM system.

Keywords: environmental sustainability; integrated environment management solutions; engineering educational program; industry–academia cooperation

Citation: Park, D.-S.; Kang, M.-S.; Chae, C.-B.; Sunwoo, Y.; Hong, K.-H. Implementation of Integrated Environmental Management and Its Specialized Engineering Education in Korea: A Case Study. *Sustainability* **2024**, *16*, 2140. https://doi.org/10.3390/su16052140

Academic Editors: Clara Viegas and Natércia Lima

Received: 24 January 2024
Revised: 22 February 2024
Accepted: 28 February 2024
Published: 5 March 2024

1. Introduction

The 1963 Act on the Prevention of Pollution in Korea initiated the resolution of environmental pollution problems. Environmental management began in an integrated manner owing to limitations of organizations and laws and was categorized according to media such as air, water quality, and waste management, as various environmental systems and laws were established, and management organizations were expanded [1]. However, the distribution of media-specific permits among pollutant-discharge facilities had various limitations, such as procedural complexity, superfluity, administrative inefficiencies, and lack of expertise [1–3]. In addition, the application of uniform emission standards, regardless of the industry, incurred social costs [4]. However, the most significant problem was that a workplace that emitted environmental pollutants could not be managed as a single pollutant source; each pollutant medium was distinguished, resulting in blind spots in pollutant and workplace management [5]. In the wake of the need for an improved environmental management system, the Ministry of Environment and various institutions that specialize in the environment in Korea worked to identify ways of resolving the problems of the system of permit- and management-specific pollutant media [6]. As a result of continuous efforts to improve this system, the Act on the Integrated Control of Pollutant-Discharging Facilities was enacted and promulgated in 2015 [7].

Integrated environmental management (IEM) is an advanced system for the comprehensive management of emission facilities, used to permit and manage pollutants individually at the workplace level [4]. In other words, it presents a paradigm shift from media-specific permits to cross-media permits. The system can aid in the transition from existing media-specific permits, thus reducing procedural complexity, and is being implemented in numerous developed countries as an advanced environmental management method [8]. As a representative example, pollutant emission facilities in the European Union (EU) have been managed using an integrated management method since the late 1990s through Integrated Pollution Prevention and Control (IPPC) and the Industrial Emissions Directive (IED) [9,10]. Following this international trend, preparations for promoting IEM have been made in Korea since 2013 [11]. However, IEM personnel in workplaces, consulting institutions, governments, and permitting institutions is lacking. In addition, working-level officials in charge of existing environmental management tasks lack expertise on the specific processes and mass balance of workplaces according to industry, and professional instructors equipped to handle all areas, such as policy, practice, materials, processes, and environmental pollution are lacking [12]. Therefore, a specialized educational project was initiated in 2020 to train experts in the IEM field in Korea in conjunction with five universities selected by the Ministry of Environment to facilitate effective management.

Therefore, the aims of this study are to examine the history and strategy of establishing an IEM policy in Korea and review essential technologies, such as emission impact analysis and Best Available Technologies (BATs), that are applied in IEM in Korea. In addition, the specialized training program for IEM in Korea and the applications of the educational system, industry–academia cooperation, and research at University K, selected as representative examples, are discussed.

2. Materials and Methods

Since the preparation for IEM began in 2013 in Korea, related laws have been investigated, referencing the system adapted by the European Union, particularly Germany, and related data have been collected [9]. After the Act on the Integrated Management of Environmental Pollution Facilities came into force in 2017, 10 environmental permits related to air, water quality, and waste under 7 statutes were combined into an integrated environmental permit system [10]. The seven statutes are the Clean Air Conservation Act, Noise and Vibration Control Act, Water Environment Conservation Act, Malodor Prevention Act, Persistent Organic Pollutant Control Act, Soil Environment Conservation Act, and Waste Management Act [11]. Having started with the electricity generation, steam, and waste treatment industries in 2017, the integrated system was applied to 19 industries (1411 worksites) by 2021. Since the Act on the Integrated Control of Pollutant-Discharging Facilities and its Enforcement Decree were established in 2017, numerous related operating guidelines and reports have been issued by the Ministry of Environment in Korea. Numerous studies on IEM are actively being conducted by various emission-source companies and institutions specializing in the environment in Korea. Therefore, in this study, various statutes, guidelines, and reference documents on Integrated Environmental Management in Korea are reviewed and synthesized to present and elucidate the system and its characteristics.

To establish these sustainability policies, programs aimed to provide IEM expertise are being developed. For solutions with engineering input, the Korean Ministry of Environment has researched education programs in cooperation with various organizations. We analyzed the human resources training programs for the sustainability market. University K, which was selected as a specialized training institution for IEM in 2020 with the support of the Ministry of Environment of Korea, is the representative case study because it has been a leader in convergence education since 2004; it established the first interdisciplinary convergence departments in Korea and has been ranked at the top since the inception of the professional IEM training program. The case study analyzes whether learners who completed the program achieved the curriculum objectives, including specialized knowledge and common skills such as writing an IEM plan and performing emission impact analysis. In addition to the educational program, an industrial cooperative project and related research were investigated. The main aspects of this study are as follows:

- Historical Background: Summary of the overall flow of the development of IEM in Korea.
- Composition of policy: Operational techniques and systems of IEM.
- The major technological elements of the policy include emission impact analysis, permit emission standards, and BATs.
- Specialized educational project: Training program to reinforce expert capacity for IEM.

This case study is an analysis of the educational program for students enrolled in the Department of Environmental Engineering and Department of Chemical Engineering at University K. In these programs, it is important to ensure that the content provided to participants is appropriately structured to suit the market's purpose [13]. To quantitatively analyze the program results, surveys as shown in Table 1 were regularly conducted, targeting training participants and field experts.

Table 1. Survey questionnaire about the program.

Survey of Participants and Experts	Program Operation Performance
• Effectiveness of education	• Number of people who completed
• Subject suitability	• Number of subjects
• Industry–academia linkage	• Number of seminars
• Job satisfaction	• Number of internships
• Instructor competency	• Research activities

3. Results

3.1. Integrated Environmental Management in Korea

IEM was first introduced in Korea in 2017 to facilitate the existing emission facility permitting systems. The system is operated in workplaces with high pollutant emissions that can cause air and water pollution [14]. The history of policies related to environmental management and the main composition and characteristics of IEM in Korea are discussed in subsequent sections.

3.1.1. Historical Background

The Prevention of Pollution Act was enacted in 1963 in Korea, and the concept of emission facilities was introduced in 1971 [15–17]. Since then, environmental issues have diversified in the 1990s, and a management system for each medium has been established through the enactment of related laws [18–20]. However, environmental management was categorized according to different media, such as air quality, water quality, and waste, as various environmental systems and laws were established, and management organizations expanded. This led to the emergence of challenges such as superfluity, procedural complexity, and inefficient management [21]. In 2003, IEM regulations were developed by the Ministry of the Environment and various environmental institutions [7,22]. Subsequently, numerous experts from related fields participated in establishing IEM policy, which was implemented in 2017 [23,24]. Figure 1 shows Korea's policy transformation history from the inception of environmental management regulation to the implementation of IEM policy.

After the implementation of IEM policy, the application period of the system was set for each industry and applied step by step for five years. In the early days of the system's operation, it targeted 19 businesses with relatively high pollutant emissions and then gradually expanded to 22 businesses [25]. Workplaces operating before the introduction of the system were incorporated into the IEM program and encouraged to obtain permits within a four-year grace period [26]. As such, the scope of workplaces managed using IEM methods is gradually expanding in Korea.

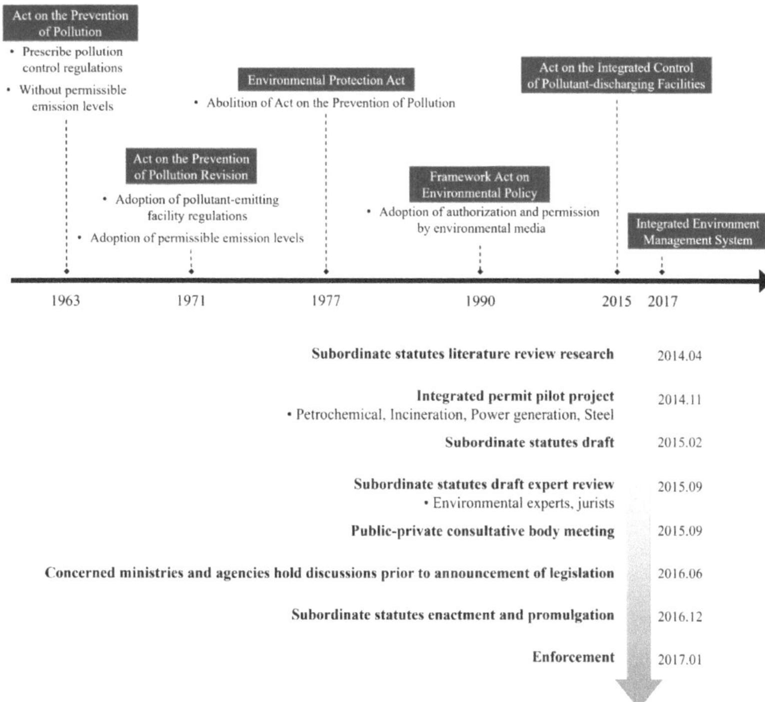

Figure 1. History of environmental management and related policies in Korea.

The implementation of IEM indicates that the environmental management paradigm in Korea has shifted from media-specific management to sources with positive effects. In the case of existing media-specific permits, uniform emission standards were set without considering the workplace features; however, IEM sets emission standards based on the pollutant-emitting industry features and workplace region [27]. Uniform permissible emissions criteria have limited impact on environmental improvement and incur unnecessary social costs; however, the implementation of IEM has resolved these shortcomings. After permission was granted, media-specific management focused on detection and crackdown was replaced with an implementation verification method focused on self-monitoring in the workplace [7]. Formerly, in the case of permit duration, once approved, the effect of a permit was maintained indefinitely, and potential problems such as the emission of new pollutants and the aging of prevention facilities owing to process changes during workplace operations could not be solved [16]. However, in the case of the IEM system, permits are reviewed every five years, which can induce the technical development of environmental management methods [28].

3.1.2. Organization and Framework of Integrated Environmental Management

IEM constitutes an integrated permit application, integrated permit review, permit confirmation procedure, and post-permit confirmation management [28]. The pollutant emission source, that is, the discharge facility, applies for the integrated permit through an Integrated Environmental Permission System (IEPS). The Specialized Environmental Assessment Institute (SEAI) then reviews the permit contents, after which they confirm whether the discharge facilities are operating according to the permit requirements based on their operation notice. After approval, the facilities continue post-management through regular inspections. Figure 2 shows the overall operating system of IEM in Korea.

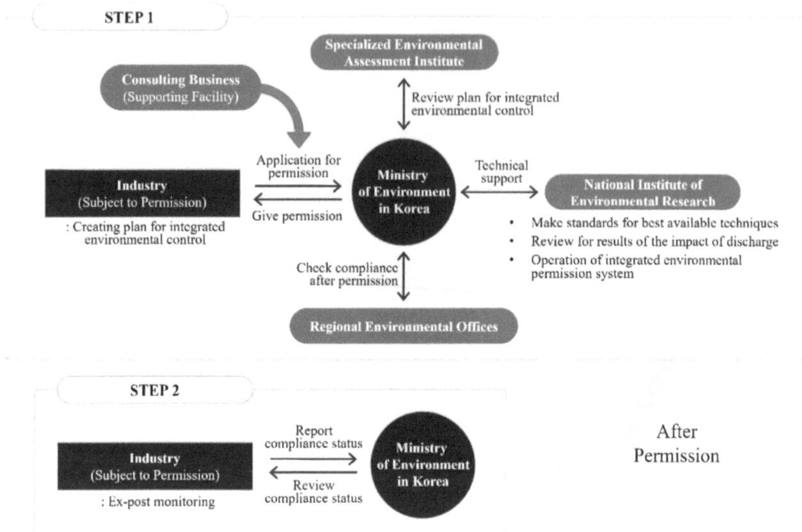

Figure 2. Integrated environmental management in Korea.

Through this process, an IEM plan constituting the main components for obtaining integrated permits is prepared and submitted. The main components, such as the establishment of permitted emission standards, integrated process charts, post-management plans, and the application of BATs, differ from those of the existing environmental permit system [29].

3.1.3. Major Techniques in Integrated Environmental Management

IEM is an advanced system aimed at resolving the problems of the existing permit system and having various scientific and technical differences from the existing method. The main technical components of IEM are as follows.

- Emission Impact Analysis and Emission Standards

In the existing permit system, uniform permissible emission criteria for each media-specific pollutant as set by the Act are applied. The emission impact analysis is performed by considering the regional and industrial characteristics of the emission sources in IEM. Whether the environmental quality meets the target level is based on the analysis results, and the emission standards that meet the target level of environmental quality in the area where the emission source is located are selected and presented as permitted emission standards [27]. To present a scientific basis for this process, the degree of pollution is calculated using an air and water emission impact analysis program, and permitted emission standards for each emission source are established within the scope of no environmental impact under environmental standards [6]. The process of establishing permitted emission standards is shown in Figure 3.

- Best Available Techniques and their Standards

The BATs represent the comprehensive environmental technologies that consider removal efficiency and economic efficiency in pollutant emission reductions [30]. The Ministry of Environment in Korea prepares and provides the BAT reference document (K-BREF) for more scientific-based IEM [31], and the BATs are applied to the integrated permission of the emission sources [32]. The K-BREF is a guideline for the efficient operation of Integrated Environmental Management. Since the implementation of the Integrated Environmental Management regulations in 2017, twenty-one references have been published in Korea.

The K-BREF is used for overall system operation, such as providing emission impact analysis guidelines and technical information on environmental management techniques for emission sources [20]. The specific application fields are illustrated in Figure 4.

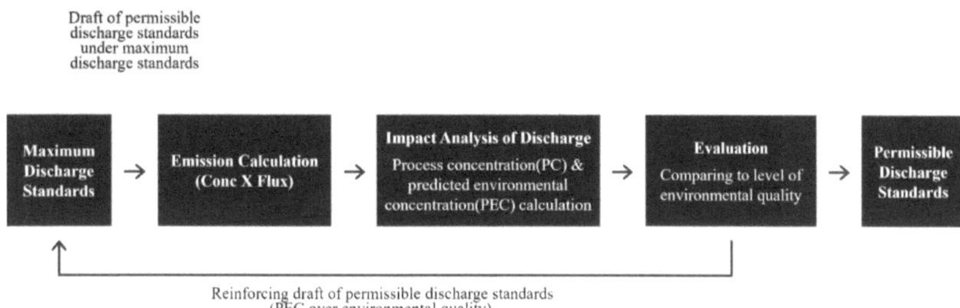

Figure 3. Process of establishing the permitted emission standards.

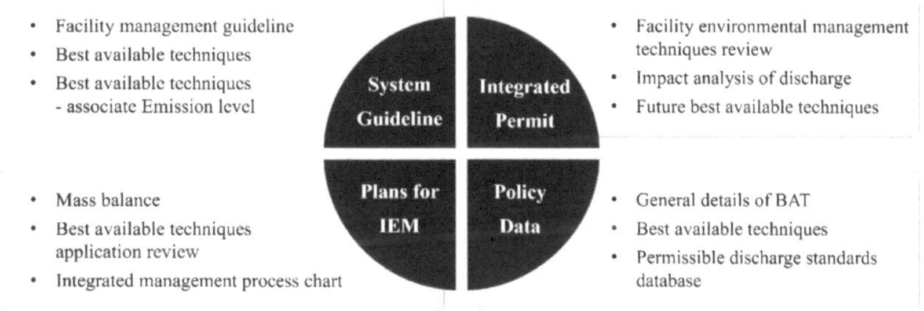

Figure 4. Availity of standards for best available techniques.

Through the application of the K-BREF in IEM, the limitations of the existing environmental improvement approach caused by uniform permissible emission criteria and social costs are addressed [33]. The lack of technical information and expertise can also be resolved through the provision of a K-BREF for each industry. This technology was developed through frequent reviews of permits and updates of the BATs, which can help solve the problems of new pollutants from process changes and old discharge prevention facilities. Consequently, it is possible to meet the demand for adequate environmental quality owing to technological development and economic development.

Table 1 shows a comparison of IEM in Europe and Korea and summarizes the system-related contents, such as the timing and scope of application. IEM in Korea has been modeled in Europe, and the system is flexibly operated with consideration of the current situation [5]. Since its introduction in 2017, the scope of its application has steadily expanded, and the system has been gradually developed [26]. In addition, K-BREFs for each industry are developed every five years. In line with the situation in Korea, where the semiconductor industry has developed, the world's first K-BREF for semiconductor manufacturing was prepared and managed [34].

3.1.4. Educational Project for Training of Professionals on Integrated Environmental Management

As mentioned above, there is a clear limit to the effect of improving environmental pollution through the existing media-specific management of emission sources. The behavior of pollutants at each stage of the emission source and the integrated tracking and management of pollutants are essential [34], but experts who can comprehensively handle these aspects are lacking. Owing to the IEM regulations, which have been in effect since 2017, the industry's demand for experts with overall knowledge of industrial processes and environmental fields has increased significantly, and education to train such experts is essential for the successful implementation of the system. Therefore, the Ministry of Environment and its affiliated organization, the Korea Environment Corporation (K-Eco), have been operating the Graduate School Project Specialized in Integrated Pollution Prevention and Control (hereafter referred to as the IEM Educational Program) since 2021.

It is important to create professional personnel who can understand the existing integrated environmental management system and provide urgent, up-to-date information. The Ministry of Environment selects and invests in excellent higher education institutions in Korea through the IEM educational program to foster experts in IEM. In particular, universities specialize in educating these engineering experts so they may be optimized for developing and researching human resource education programs. The five universities selected through strict evaluation are supported by a total budget of more than USD 7 million (KRW 9 billion) and produce approximately 100 people with specialized training in IEM every year. Each university operates a specialized graduate school aimed at improving the technical skills of practical-oriented professionals in IEM through the systematic development of educational programs and research in connection with industries. The curriculum includes preparing an IEM plan, which is essential for applying for integrated permits, and major core common subjects, including the EIA. Additionally, each university conducts a convergence course between related majors and forms a joint industrial practice (consortium) curriculum.

3.2. Case Study: IEM Educational Program at University K, Seoul

A case study was conducted to better understand the programs of graduate schools specializing in integrated pollution prevention and control as promoted by the Ministry of Environment in Korea. University K, which was evaluated as the best case among the institutions selected for such programs, was selected and investigated as an expert training program for IEM. University K developed a human resources training program as a way to address integrated environmental management problems; analysis of this case can be useful for future applications in the field of sustainability.

The IEM Educational Program at University K has been in operation since 2021 and consists of an educational curriculum, research, and industry–academia cooperation, as shown in Figure 5. University K serves as a hub between the Ministry of Environment, policymakers, and the industries that implement the system, forming a positive feedback structure. To cultivate the experts required for the operation of the IEM policy, opinions from the government were collected, and various related industries and consortia were formed and exchanged to produce professionals with the necessary knowledge in the industry. To complete this program, it is essential to take courses related to IEM, participate in industry–academia cooperation programs, and conduct research. Students with sufficient understanding of IEM systems through the completion of the professional curriculum are immediately placed in the industrial field to meet the demand for professional experts in the industry.

3.2.1. Curriculum

With the implementation of IEM policies by industries, the need to cultivate professionals who can comprehensively understand pollutant tracking has increased. Therefore, the IMG Educational Program conducted by University K has organized and operated a professional curriculum to train experts. The curriculum consisted of subjects for the preparation of an Integrated Environmental Management plan, emission impact analysis, calculation of mass balance, and understanding of the process, including emission and prevention facilities.

As presented in Figure 6, University K established an IEM educational program management committee to continuously discuss improvements in the curriculum and manage students in the program. In addition, demand surveys of industries are periodically conducted to reflect the opinions of practitioners, and the curriculum is improved and reorganized through the resolutions of a regular steering committee meeting twice a year.

The curriculum is divided into three courses based on the importance of each subject: fundamental, advanced, and applied. The compositional subjects cover the overall content and scope, including the basics related to Integrated Environmental Management and the media's understanding of management and systems. In addition, the curriculum includes engineering undergraduate courses related to Integrated Environmental Management to expand opportunities for undergraduate students to participate in the program.

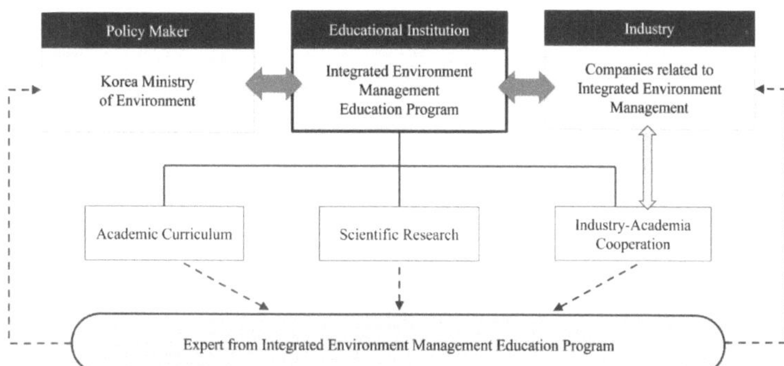

Figure 5. Overview of integrated environment management education program at University K.

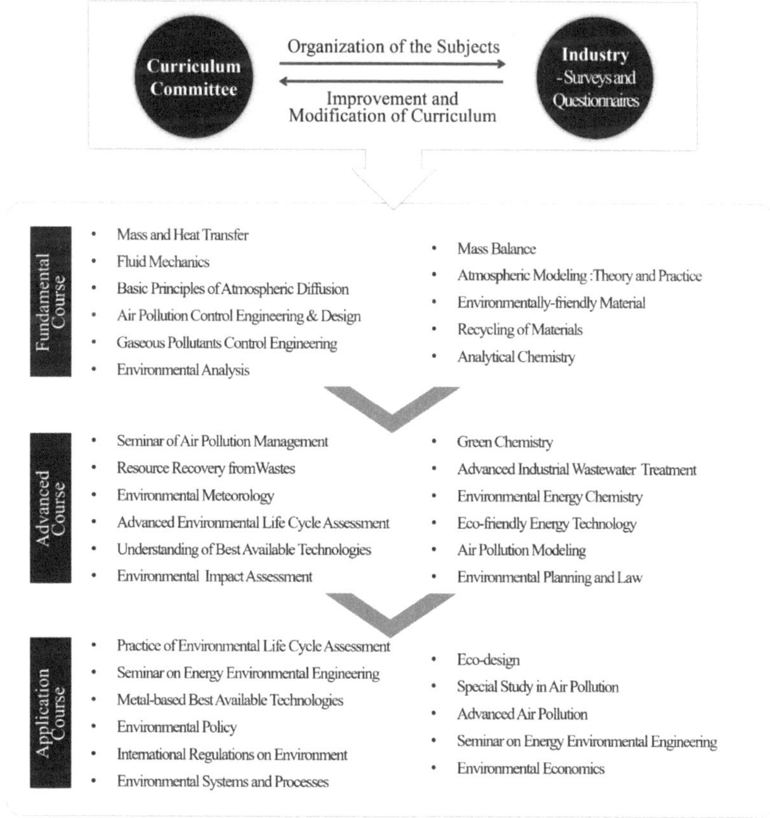

Figure 6. Curriculum organization and process by committee.

3.2.2. Industrial Cooperative Project

Close cooperation with industries is essential to improving the employment competitiveness of students who have completed the program and meet the requirements for education in the field. University K formed a consortium with major industries in the IEM field to provide students with education at industrial sites. The IEM Educational Program has established industry–academia partnerships with Integrated Environmental Management consulting industries, pollutant emission sources, governments, and numerous research institutes. Through these connections, opinions from various field experts are collected, which contribute to the development of the program and curriculum.

More than 15 industry–academia cooperation programs (seminars, field practices, and internships) are organized annually by the IEM Educational Program at University K to provide practical education. Seminars and field practices for each semester are planned for each main topic, as shown in Table 2, and the field practices are organized based on the established industry–academia partnership. The on-site workshop provides opportunities to learn from experts and acquire related legal instruction from national and public research institutes responsible for policy implementation, as well as information from the perspectives of IEM consulting companies. Students participating in the program acquire knowledge and skills related to the IEM system from a pool of experts with various perspectives.

Table 2. Integrated environment management in Europe and Korea.

	Europe	Korea
History	• 1996, European commission's integrated pollution prevention and control directive • 2011, Industrial emission directive	• 2015, Act on the Integrated Control of Pollutant-Discharging Facilities • 2017, Integrated environment management system
Range	• 33 types of business	• 22 types of business
Integrated pollutants	• Determined according to the circumstances of each country.	• Air pollutants, volatile organic compounds, fugitive dust, noise, vibration, water pollutants, malodor, persistent organic pollutants, soil contaminants, wastes
Discharge facility management after permission	• Environmental inspections, Inspection of compliance with permission details • Emissions monitoring	• After permission report • Discharging facility monitoring • Annual IEM reports
BAT management	• BREFs (BAT reference documents) • Periodic update (3 years)	• K-BREFs • Periodic update (5 years)

3.2.3. Research

In the IEM Educational Program at University K, major topics related to the aforementioned educational curriculum were selected, and in-depth research was conducted separately. Students participating in the program conduct various subject-specific application studies, such as modeling the impact assessment of pollutants, process-specific BATs, and life cycle assessment (LCA)-based mass balance based on an understanding of the IEM system. Table 3 shows the research content based on key topics. Also, Table 4 shows major research projects in IEM program.

Table 3. Main topics of industrial cooperative program.

Main Topic	Details
(1) Theoretical education on integrated environmental management system	- Education on laws related to integrated management of pollution facilities - Education on the relationship and differences between the integrated law and existing laws
(2) Basic education on integrated permit consulting duties	- Learning technical theories such as BATs (best available techniques) - Learning how to conduct various programs such as impact analysis of pollutants
(3) Practical education on integrated permits	- Education on how to write an integrated environmental management plan - Understanding the integrated permit process and precautions - Understanding how to use the integrated permit system
(4) Education on the ability to promote integrated permits	- Developing the ability to analyze environmental issues and propose solutions - Improving the ability to reach agreements through discussions

Table 4. Research projects in IEM program.

Key Topic	Summary of Research
Improvement of material balance calculation methodology based on O-LCA	The aim is to prepare an O-LCA-based material balance calculation methodology by analyzing the most basic standards dealing with product LCA, ISO 14040:2006 [35], 14044:2006 [36] and KS I ISO/TS 14072:2014 [37] as well as Guidance on Organizational Life Cycle Assessment. At workplaces that implement the integrated environmental management system, it is possible to identify the causal relationships between inputs and outputs that were not identified when dealing with individual media and to identify abnormal flows in processes and facilities.
Improvement of pollutant reduction technology through customized convergence technology for emission sites	The purpose is to propose and present application plans for optimal utilization techniques, which are common points of the total amount of pollutants and the integrated environmental management system. This study facilitates establishment of a comprehensive evaluation model of quantitative and qualitative evaluation factors applied for integrated environmental management by pollutant emission type. In addition, the impact assessment related to the technical excellence, environmental management performance, economic feasibility, maturity of operational and management technologies, and sociality (local feasibility) of each evaluation factor is conducted.
Utilization and improvement of integrated environmental permit air and water pollution impact analysis program	This is a simulation program to investigate and analyze the impact of pollutant emissions from workplaces on the surrounding environment and is used to set customized permit emission standards. It considers existing pollution levels and additional pollution levels and involves conducting research on improvement measures by identifying existing problems.

Students participating in the program undertake various educational projects, including in-depth research, as described above. In other words, this program aims to educate and train experts with a wide range of relevant theories and practical experience in IEM to meet the demands of the actual field.

4. Conclusions

This study examined and discussed the introduction and development of IEM policy in Korea, along with its operational system and established technologies. Following the introduction of IEM, the environmental management approach in Korea shifted towards pollutant source management. This has led to positive outcomes such as the establishment of a scientific basis for emission regulations, self-management by businesses, and gradual reinforcement of emission regulations through periodic permit reviews. In addition, the introduction of emission impact analysis and BAT criteria, the main technical components of IEM, provides a scientific basis for system operation and addressed the lack of expertise in managing businesses based on existing media-specific approaches.

The specialization of the personnel in charge is essential to increase the completeness and implementation of Integrated Environmental Management. Specialized education programs to train experts in the field of Integrated Environmental are led by the Korean government. To study the educational program model, University K, the best of five institutions selected by the Ministry of Environment, was chosen for analysis. As investigated in this study, the expertise needed in the current sustainability market must include diverse experiences and theoretical knowledge. In order to analyze the goals of the integrated environmental management system, well-constructed educational cases were selected and analyzed. Through regular surveys and tracking, engineering students who completed the above education program were confirmed to meet the qualifications for sustainability development challenges. The seminars, field practice, and research activities for specialized subjects in the program were carried out through close cooperation with various related institutions, and the quality of education was greatly improved. However, the period in which the system was implemented was relatively short, so related information is limited. Therefore, it is important to collect data from the beginning and accumulate feedback on any problems. Also, because policies are continuously expanding, research in new areas is also necessary. This educational model can provide guidelines for countries preparing to introduce Integrated Environmental Management in the future. In particular, the goals and planned composition proposed in this program are believed to be of great help in the design and strategic development of the curriculum for fostering expertise.

Author Contributions: Conceptualization, Y.S. and K.-H.H.; methodology, M.-S.K.; software, D.-S.P.; validation, K.-H.H., Y.S. and D.-S.P.; formal analysis, D.-S.P.; investigation, M.-S.K.; resources, C.-B.C.; data curation, D.-S.P.; writing—original draft preparation, D.-S.P.; writing—review and editing, K.-H.H.; visualization, D.-S.P.; supervision, K.-H.H.; project administration, Y.S.; funding acquisition, Y.S. All authors have read and agreed to the published version of the manuscript.

Funding: This work was financially supported by the Korean Ministry of Environment (MOE) Graduate School specializing in Integrated Pollution Prevention and Control Project.

Institutional Review Board Statement: Not applicable.

Informed Consent Statement: Not applicable.

Data Availability Statement: Data is contained within the article.

Acknowledgments: We want to express our sincere gratitude to the anonymous reviewers and editors for their efforts in improving the paper.

Conflicts of Interest: The authors declare no conflicts of interest.

References

1. Park, J.H.; Kim, Y.S.; Lee, W.S. Significance and Development of the Integrated Environment Management System Implementation. *J. Korean Soc. Water Environ.* **2016**, *32*, 318–324. [CrossRef]
2. Environmental Permitting Guidance: The Large Combustion Plants Directive. Available online: https://www.gov.uk/government/publications/environmental-permitting-guidance-the-large-combustion-plants-directive (accessed on 26 December 2023).
3. Korean Society on Water Environment; National Institute of Environmental Research. *Enhanced Integration Prepared Based Integrated Environmental Management and Environmental Management of Water*; National Institute of Environmental Research: Incheon, Republic of Korea, 2013.

4. Han, D.H.; Yang, I.J.; Kang, P.G.; Kim, E.S. Policy Research Direction for Integrated Environment Management. *Proc. Korea Air Pollut. Res. Assoc. Conf.* **2019**, *62*, 49.
5. Shin, Y.S. Review on the Current Status of Integrated Multimedia Risk Assessment for Implementing Integrated Environmental Management in Korea. *Environ. Law Policy* **2009**, *2*, 145–166.
6. Seo, J.H.; Hwang, H.-J.; Khan, J.; Lee, S.; Kim, Y. The Present Status and Improvement of Air Quality Assessment Methodology for Integrated Environmental Permission System. *J. Korean Soc. Atmos. Environ.* **2021**, *37*, 523–535. [CrossRef]
7. Kim, B.G. Development of the Integrated Environmental Management Policy. *Proc. Korea Air Pollut. Res. Assoc. Conf.* **2019**, *62*, 47.
8. Kim, E.; Kim, G.; Seo, K.; Khan, J.-B.; Seok, H.; Kim, Y.; Kang, P.-G. Introduction on Best Available Techniques Reference Document for the Display Manufacturing Industry in Korea. *Kihm* **2019**, *7*, 42–49. [CrossRef]
9. Jeong, H.-Y. EU Envioronmental Policy and Their Impact on Korean Industries. *J. Contemp. Eur. Stud.* **2007**, *24*, 263–302. [CrossRef]
10. Lee, H.J. The Expected Roles of Environmental Consulting Company for the Integrated Environment Permitting Scheme. *Proc. Korea Air Pollut. Res. Assoc. Conf.* **2019**, *64*, 48.
11. Seo, K.; Kim, E.; Kim, G.; Khan, J.; Hong, S.; Kang, P. Understanding and Improvement of the K-BREF (Korea BAT Reference Documents) for the Corrugated Cardboard Manufacturing Industry. *J. Environ. Sci. Int.* **2020**, *29*, 559–573. [CrossRef]
12. Park, J.-H.; Seok, H.-J.; Kang, P.-G.; Ahn, H. Utilization of an Information System for the Efficient Implementation of the Integrated Environmental Permit System in South Korea. *Sustainability* **2023**, *15*, 16512. [CrossRef]
13. Zhang, Y.; Chen, X. Empirical Analysis of University–Industry Collaboration in Postgraduate Education: A Case Study of Chinese Universities of Applied Sciences. *Sustainability* **2023**, *15*, 6252. [CrossRef]
14. Song, J.H.; Jang, H.S. Problems and Discussions on the Korean Integrated Pollution Prevention and Management Law Cases in Air Pollution and Odor. *Proc. Korea Air Pollut. Res. Assoc. Conf.* **2019**, *62*, 49.
15. Framework Act on Environmental Policy. Available online: https://elaw.klri.re.kr/kor_service/lawView.do?hseq=60971&lang=ENG (accessed on 26 December 2023).
16. Park, J.H.; Shin, S.J.; Lee, D.G. Enhancing the Applicability and Improvement Direction of Integrated Environmental Permit System. *J. Korean Soc. Water Environ.* **2018**, *34*, 339–345. [CrossRef]
17. Lee, J.H. Integrated Environmental Management and Environmental Impact Assessment. *J. Korean Cadastre Inf. Assoc.* **2018**, *20*, 89–103. [CrossRef]
18. Clean Air Conservation Act. Available online: https://elaw.klri.re.kr/kor_service/lawView.do?hseq=57423&lang=ENG (accessed on 26 December 2023).
19. Water Environment Conservation Act. Available online: https://elaw.klri.re.kr/kor_service/lawView.do?hseq=54838&lang=ENG (accessed on 26 December 2023).
20. Beak, S.J.; Lee, S.W.; Yun, Y.B.; Kim, D.C. The BAT Reference Document for Enforcing the Integrated Environmental Management System. *Proc. Korea Air Pollut. Res. Assoc. Conf.* **2014**, *57*, 85.
21. Kim, D.Y.; Cho, Y.M.; Choi, M.A.; Jun, S.Y.; Hong, L.S. *A Policy Implication of Integrated Pollution Prevention and Control Program in Gyeonggi-Do*; Policy Research; Gyeonggi Research Institute: Suwon, Republic of Korea, 2014; pp. 1–53.
22. Lee, S.K. Conceptual Framework of Integrated Environmental Management. *Korean Soc. Public Adm.* **2004**, *15*, 313–332.
23. Act on the Integrated Control of Pollutant-Discharging Facilities. Available online: https://elaw.klri.re.kr/kor_service/lawView.do?hseq=55955&lang=ENG (accessed on 26 December 2023).
24. Kim, D. Research on Improvement of Permission System under Act on the Integrated Control of Pollutant-Discharging Facilities. *Environ. Law Rev.* **2020**, *42*, 57–87. [CrossRef]
25. Enforcement Decree of the Act on the Integrated Control of Pollutant-Discharging Facilities. Available online: https://elaw.klri.re.kr/eng_mobile/viewer.do?hseq=55955&type=part&key=39 (accessed on 26 December 2023).
26. Kim, J.H.; Jeong, S.Y.; Na, M.H.; Choi, J.H.; Im, C.I. A Case Study and Review of Emission Impact Analysis for Establishing Integrated Environmental Management Permit Emission for Air Pollutants-Emitting Facilities. In Proceedings of the Korea Environmental Policy and Administration Society Conference, Gwangju, Republic of Korea, 23 June 2017; pp. 9–23.
27. Seok, H.-J.; Kahn, J.-B.; Kim, Y.-L.; Seo, J.-W.; Hong, S.-Y.; Kim, H.-J. Discharge Impact Analysis of Air Pollutants for Integrated Environmental Management. *J. Environ. Anal. Health Toxicol.* **2020**, *23*, 240–247. [CrossRef]
28. Kim, J.H.; Ha, T.Y.; Na, M.H.; Lee, J.H. A Case Study on Integration Permitting of Power Generation Sector by Implementation of Integrated Environmental Management System. *Korean Policy Sci. Rev.* **2018**, 27–30. [CrossRef]
29. Guem, H.S. Introduction of an Integrated Approach into the Environmental Permitting and Management System. *Proc. Korea Air Pollut. Res. Assoc. Conf.* **2014**, *57*, 85.
30. National Institute of Environmental Research. *Plan for Integrated Environmental Control Guidelines*; Ministry of Environment in Korea: Sejong, Republic of Korea, 2020.
31. National Institute of Environmental Research. *Best Available Techniques Reference of Korea—Plastic Product Manufacturing Industry*; Ministry of Environment in Korea: Sejong, Republic of Korea, 2020.
32. European Union (EU). *Directive 2010/75/EU of the European Parliament and of the Council of 24 November 2010 on Industrial Emissions*; Integrated Pollution Prevention and Control: Dresden, Germany, 2010.
33. National Institute of Environmental Research. *Best Available Techniques Reference of Korea—Steel Manufacturing Industry*; Ministry of Environment in Korea: Sejong, Republic of Korea, 2017.

34. Baek, J.; Cho, K. A Study on the Factors Influencing Regulatory Awareness: A Focus on the Sites Subject to the First Year of the Integrated Environmental Management System. *Kpsr* **2020**, *24*, 105–129. [CrossRef]
35. *ISO 14040:2006*; Environmental Management—Life Cycle Assessment—Principles and Framework. ISO: Geneva, Switzerland, 2006.
36. *ISO 14044:2006*; Environmental Management—Life Cycle Assessment—Requirements and Guidelines. ISO: Geneva, Switzerland, 2006.
37. *ISO/TS 14072:2014*; Environmental Management—Life Cycle Assessment—Requirements and Guidelines for Organizational Life Cycle Assessment. ISO: Geneva, Switzerland, 2014.

 *sustainability*

Article

Implementation of Environmental Engineering Clinics: A Proposal for an Active Learning Methodology for Undergraduate Students

Dante Rodríguez-Luna [1], Olga Rubilar [2], Marysol Alvear [3], Joelis Vera [4] and Marcia Zambrano Riquelme [2,*]

1 Facultad de Ingeniería, Instituto de Ciencias Químicas Aplicadas, Universidad Autónoma de Chile, Santiago 8910060, Chile; derodriguez@alu.ucam.edu
2 Departamento de Ingeniería Química, Universidad de La Frontera, Temuco 4811230, Chile; olga.rubilar@ufrontera.cl
3 Departamento de Ciencias Químicas y Recursos Naturales, Universidad de La Frontera, Temuco 4811230, Chile; marysol.alvear@ufrontera.cl
4 Facultad de Ingeniería y Ciencias, Universidad de La Frontera, Temuco 4811230, Chile; j.vera12@ufromail.cl
* Correspondence: marcia.zambrano@ufrontera.cl; Tel.: +56-45-232-5472

Abstract: Quality education focused on quality, inclusion, and opportunity is one of the United Nations Sustainable Development Goals to reduce inequality in the knowledge of the people who are educated. In this sense, universities have a role in rethinking the teaching model, changing their strategies, and including new experiences based on active learning. This article makes a didactic methodological proposal for undergraduate and graduate students using learning experiences for solving regional environmental problems proposed by municipalities. This method considered creating an agreement, defining topics, preparing bases and study areas, analyzing problem solutions, and delivering products. The results showed the implementation of the environmental engineering clinics (ECCs) in five subjects of the curriculum, with the participation of sixty students, who solved problems from seven municipalities. The results showed a correct implementation of the active learning methodology, allowing for knowledge to be transferred in a real-life scenario, significantly facilitating student learning. The plan–do–check–act (PDCA) cycle provides a practical framework for learning while solving real-world challenges, empowering learners to personally engage with authentic and meaningful challenges within their communities. As was previously stated, this article presents a methodology that can be introduced in universities to improve the learning process through active learning and the link with real problems of the territories where they are located, which also allows for improving the connection with the environment, contributing significantly to the sustainability of the territories.

Keywords: engineering education; empirical learning; environmental engineering; Chile

Citation: Rodríguez-Luna, D.; Rubilar, O.; Alvear, M.; Vera, J.; Zambrano Riquelme, M. Implementation of Environmental Engineering Clinics: A Proposal for an Active Learning Methodology for Undergraduate Students. *Sustainability* **2024**, *16*, 365. https://doi.org/10.3390/su16010365

Academic Editors: Clara Viegas and Natércia Lima

Received: 15 October 2023
Revised: 21 December 2023
Accepted: 27 December 2023
Published: 30 December 2023

1. Introduction

The Sustainable Development Goals (SDGs) established goal 4 to provide "quality education", considering quality and inclusion as the main focuses, and must be accompanied by different opportunities for better learning [1]. In this sense, the fulfillment of goal 4 makes it possible to reduce the gap in the knowledge of the people who are educated, and this contributes, in the long term, to the realization of the other goals [2].

Sustainability must be understood as a balance between social, economic, and environmental factors. For this, it is necessary to teach students engineering and forms of relationship and practical experiences. This has implied a rethinking of the teaching model, which, until then, was based on teacher-based teaching, transferring this new approach into an active learning model based on strategies that focus learning on the student, not turning the system around the teacher, but rather a function of the activities carried out by the student to achieve their learning objectives [3]. The implementation of an active learning program has been recommended by professional European engineering associations and the UNESCO Active Learning in Engineering Education network, among others [4,5].

The change in roles between students and teachers must be reflected in teaching planning [3]. Likewise, there is evidence that case problem analyses and resolution methodologies in groups with few students favor learning [6]. In this sense, it is a priority to rethink the traditional teaching model, migrating from teaching centered on the teacher to one centered on the student, considering that teaching practices are decisive in student learning [3,7]. This is a complex process, since it requires modifying the study plans and, in general, introducing methodological innovations in a traditional teaching model, in addition to the active and deep participation of the students in developing the activities. Focusing the learning process on the student implies a teaching effort since it requires being the conductor of the orchestra, that is, directing, but not conducting, but instead teaching how to learn by performing [8].

On the other hand, engineering students who are essential for meeting this SDG [9] go from being passive recipients of knowledge or information to active participants in the learning process, sharing, reflecting, and deciding the best strategy to address the problem guided by their teacher [8]. Active learning considers several active methodologies: analysis and problem-based learning, team learning, and flipped classrooms [6,10–12]. Problem-based studies are undoubtedly one of the most relevant strategies, given the integration of theoretical knowledge, its applicability to real situations, and its integration [10,13]. In this sense, it is necessary to consider that students improve their learning process by actively participating in the experience, creating everyday situations in real-life activities [14,15].

In this sense, Vygotsky's theory focused on social development becoming relevant, which states that social interactions are essential to people's cognitive development and learning. Figure 1 shows the main aspects of the sociocultural theory of Vygotsky [16,17] that has been oriented to the teaching role to maximize learning effectiveness.

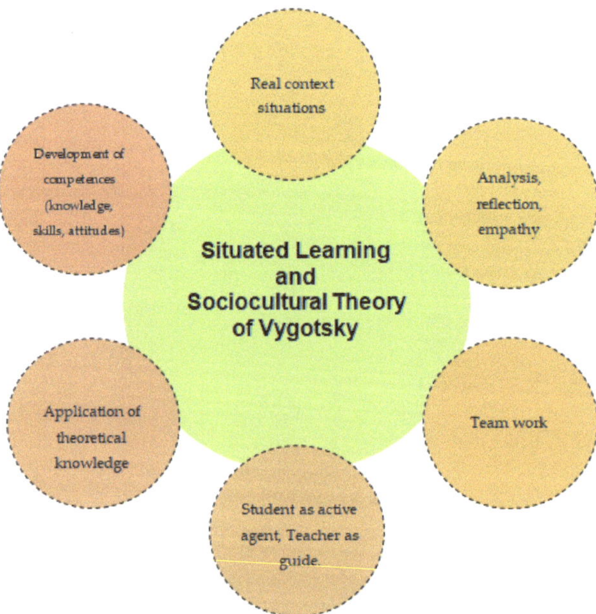

Figure 1. Conceptual or theoretical scope of the innovation experience based from Vygotsky's theory [16,17].

In Vygotsky's theory [16], the realization of activities focused on sociocultural aspects can improve the skills of the students and even their cognitive development. For the implementation of this theory, the role of the teacher is essential, since this teaching-learning methodology seeks to generate solutions, generating significant learning while searching for alternative solutions. On the other hand, the teacher has a vital role since they must stimulate the critical thinking of the group and permanently guide it [18].

In the global context, the environmental engineering clinics (ECCs) arose from the need to solve environmental problems for which there were no economic or human resources in the municipalities to be able to solve them. Considering the above, the ECC has emerged as a methodology that allows for students of the environmental civil engineering degree to be linked early with the environmental problems of the region, that is, problems that they will encounter when they enter the world of work. The main structure of the ECC is focused on the student (student centered). The continuous improvement of community problems is based on plan–do–check–act (PDCA) cycles supported by standard work concepts. Every cycle is planned and executed, and then the results are verified with the commune user, and decisions on the next cycle are taken [19,20]. Environmental civil engineers have a fundamental role in promoting sustainability based on their acquired knowledge and convictions. The ECCs contribute significantly to sustainability, as they consider the three basic elements of sustainability and were conceived to search for environmental solutions that are related to social demands in contexts where economic resources do not exist to address the environmental problems that are identified in the territories.

This article provides a didactic methodological proposal to promote the professional training of undergraduate and postgraduate students through situated learning experiences through the resolution of regional problems on environmental issues. Thus, it has been possible to deliver tools and information to the municipalities of the Araucanía region that facilitate their local environmental management through the action developed by the students through the ECCs.

This model includes a manual of procedures based on the application of situated and experiential teaching, facilitating active learning focused on meaningful experiences. In this way, students have been linked from the first to the last year with real problems, which allows for them to develop their talents, put them into practice, and install social responsibility with activities of this sort. Empathy is the crucial factor that enables this methodological approach to be successful, since it provides the necessary approach to real problems from the users to the student teams to search, study, formulate, and create solutions.

2. Methodology

2.1. Agreement Creation

First, an agreement was signed between the Universidad de La Frontera (UFRO) and the Ministerio del Medio Ambiente (MMA), which provided the formal basis for establishing a relationship of coordination and collaboration between these counterparts, designating professionals as facilitators that were responsible for the development of activities around the resolution of regional problems in environmental matters, with a particular emphasis on the municipalities of the Araucanía region, belonging to the Sistema de Certificación Ambiental Municipal (SCAM) [21]. Chile is located in South America, has 17.5 million inhabitants, and a length of 4,200 km of continental territory. The Araucanía region is located in the South of the country and has 32 municipalities [22,23].

2.2. Participants and Definition of Subjects

At the beginning of each academic semester, a call was made to the different professors of the mandatory subjects of the Faculty of Engineering and Sciences to participate in the EECs. Once the issues were confirmed, the UFRO and the MMA analyzed and discussed the learning opportunities to be addressed that were inherent to the professional profile and consistent with the environmental problems of the municipalities in the Araucanía region [21]. Consequently, all students (participants) in the subjects that were selected had to essentially participate in the ECCs. The participants were aged between 17 and 23 years old and were students of environmental civil engineering between the 1st and 11th academic semesters.

2.3. Preparation of Bases and Study Areas

Then, the bases were prepared, and the following were established:

- Subjects available for case studies.
- Student competencies categorized by level and learning outcomes.
- Essential and mandatory requirements for municipalities, e.g., attendance at meetings, provision of data for studies, etc.
- Obligations of the university, e.g., analyze the cases under a PhD-holding individual's supervision and deliver a final report with the results to the municipality.
- Selection criteria for selecting the cases: clarity of the requirements (15%), technical relevance (25%), consistency of the request with the environmental strategy of the city (30%), coherence of the request with the regional problem (10%), and contribution from the municipal counterpart (20%).
- Members of the selection committee: A representative from the UFRO and another from the MMA are appointed. The selection committee reviews the applications and their correspondence with the requirements established in the bases and based on the score obtained, the case studies are selected. Then, an adjudication document is prepared and sent to all the applicant municipalities.
- Deadline for applications: the application deadlines are established, an application form is delivered, and the e-mail to which the application should be sent is indicated.

Then, when the bases were completed, they were formally sent by the MMA to the municipalities of the Araucanía region attached to the SCAM.

2.4. Problem Solutions and Product Deliveries

Once the cases under study were awarded, a meeting (in person or online) was coordinated between the teacher and students of the subject and the municipal counterpart indicated in the application form. During this meeting, detailed information on the problem was delivered, and subsequent instances were coordinated to review the joint progress of the case under study between the parties [21].

The teaching strategies of the subjects linked to this project were based on situated and experiential teaching, facilitating active learning focused on meaningful and motivating experiences (authentic), and promoting critical thinking and awareness.

The intermediate results of this process were presented internally in instances according to the subjects involved, and the final presentation was made in some cases in the respective municipalities and at the closing ceremony of the EEC, one week before the end of the academic semester. A public presentation of the results was carried out, and a document with these results was delivered to the person in charge of the municipality. In this presentation, two members of each subject made a brief technical presentation of the results of the problem addressed in the EEC.

During this stage, the selection committee follows up by organizing at least three meetings (throughout the semester) with the academics responsible for the subjects to monitor the progress of the activities of the requirements received in the EEC, guaranteeing the product's delivery during the closing ceremony.

Figure 2 summarizes the stages of this process from the preparation of the bases to the delivery of the product.

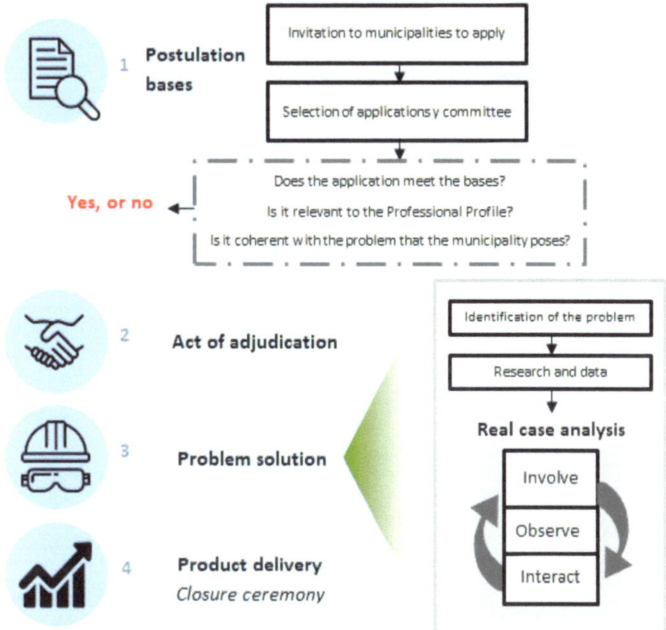

Figure 2. Stages of the process.

3. Results

3.1. Definition of Subjects

In the meeting that was called with all the teachers of the different subjects of the Faculty of Engineering and Sciences, the options for implementing the ECC in the academic semester were analyzed. This analysis considered the learning objectives of the subject, the possibility of incorporating practical experiences, and the teacher's interest in innovating through the ECC. The implementation of the ECC agreed upon jointly with the teachers is shown in Table 1.

Table 1. Subjects that were selected to implement ECCs during a semester.

N°	Subject	Semester	Number of Enrolled Students
1	Environmental analytical chemistry	7th	5
2	Wastewater engineering	8th	14
3	Industrial processes	3rd	11
4	Solid waste engineering	9th	16
5	Engineering project	11th	14

All of the above subjects belong to university studies on environmental civil engineering and are taught by professors with a PhD. The selected subjects are mandatory and are located between the 3rd and 11th semesters, and the number of students fluctuates between five and sixteen. It should be noted that environmental civil engineering has a duration of 12 academic semesters, and all students enrolled in the subjects must participate by solving the selected problem.

3.2. Applicant Municipalities

Thirty-two municipalities in the Araucanía region were invited to participate in the ECC. In the call, the requirements and selection criteria established for the selection were announced. The following table (Table 2) shows the selected municipalities and associated subjects, all being municipalities of the Araucanía region.

Table 2. Municipalities that were selected and problems or needs raised.

ID Municipality	Characteristics	Population (Inhabitants) [22]	Problem or Need Raised	Subject
A	Located in a mountainous zone	10,251	Sampling and physical–chemical characterization of the San Pedro and Jara lagoons	
B	Located in the central valley	34,182	Sampling and physical–chemical characterization of the Traiguén river	Environmental analytical chemistry
C	Located in the central valley	6905	Sampling and physical–chemical and biological characterization of the wastewater treatment plant and Perquenco estuary.	
D	Situated in a coastal zone	12,450	Sampling and physical–chemical characterization of the "May 21 estuary"	
E	Located in a central valley municipality	38,013	Diagnosis and recommendations to improve the operation of the Lautaro fish farm.	Wastewater engineering
D	Situated in a coastal area	12,450	Analysis of alternatives for wastewater treatment in rural areas.	
D	Situated in a coastal area	12,450	Training in sustainable entrepreneurship	Industrial processes
F	Situated in a coastal area	32,510	Analysis of alternatives for the treatment of city waste	Solid waste engineering
G	Located in a north zone	7733	Design of a composting plant for the city of Ercilla	Engineering project

In this first version of the ECC, three applications from municipalities were rejected, mainly because the requests were not related to the learning objectives of the subjects, the problem was not well identified, and/or the expected result was not specified. In general, there was great interest on the part of the municipalities in applying.

3.3. Products Obtained by Subject

At the global level, the students applied the plan–do–check–act (PDCA) cycle, which is a four-step model for carrying out changes. The PDCA cycle is considered a project planning tool for process improvement that focuses on continuous learning and knowledge creation and is undertaken as a cycle [24,25]. Figure 3 shows an approach with a "plan" step to meet the requirements of the municipalities to provide answers that adjust to their needs of real problems.

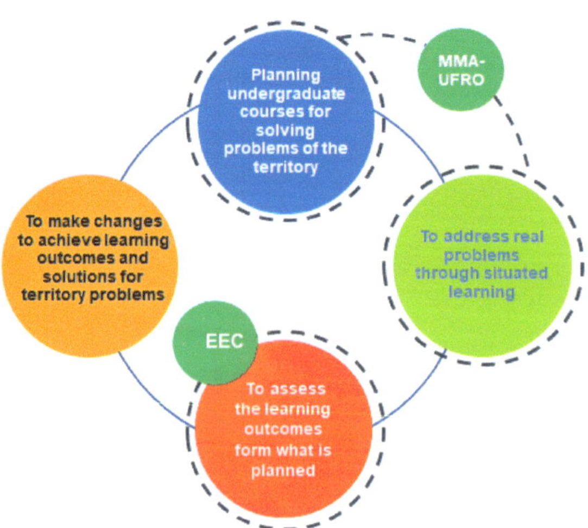

Figure 3. Continuous loop of planning, doing, checking, and acting.

The results that were obtained in each of the subjects are detailed below.

3.3.1. Environmental Analytical Chemistry

The five students, under the supervision of a teacher, took samples and characterized the water bodies of the four cities that requested them (Figure 4). The determination of water quality in lagoons, rivers, and estuaries was relevant for these municipalities; in fact, this was the topic of greatest demand. The results that were obtained are shown in Table 3. These results were compared with several parameters of the Chilean Standard (NCh) 1333, "Water quality requirements for different uses". This Chilean standard establishes that with a conductivity of less than 750 µS/cm, no harmful effects are observed for the irrigation of plants [26].

Figure 4. Sampling process in municipality 1.

Table 3. Laboratory results that were obtained.

Parameter	Unit	San Pedro Lagoon	Jara Lagoon	Traiguén River	Perquenco Estuary	May 21 Estuary
pH	-	6.4	6.9	6.3	6.67	7.0
Conductivity	µS/cm	148.1	70.4	53	78	507
Turbidity	NTU	3.2	13.0	19	6.55	28
Settleable solids	mL/Lh	<0.2	5	<0.2	0.2	<0.2
Total solids	mg/L	286	236	250	-	460
Total suspended solids	mg/L	158	160	114	3.8	198
Total phosphorus	mg/L	<0.5	1.38	<0.5	<0.5	-
Ammoniacal nitrogen	mg/L	<1	<1	<1	<1	<1
Total coliforms	NMP/100 mL	2.3×10^2	7.9×10^2	-	7.9×10^3	1.1×10^3
Fecal coliforms	NMP/100 mL	<2	70	-	2.3×10^3	1.1×10^5

In the same way, NCh 1333 showed that for recreation with direct contact, the pH must be in the range between 6.5 and 8.3, the turbidity must not exceed 50 NTU, and they must have a maximum of 1000 NMP/100 mL for fecal coliforms. The requirements of NCh 1333 also establish general conditions for waters intended for aquatic life, which indicates that the pH must be between 6.0 and 9.0, and the settleable solids must not exceed the natural value of waters designed for aquatic life. Therefore, the only parameters that exceeded the regulations were the fecal coliforms in Perquenco estuary and May 21 estuary, which indicate microbiological contamination in their waters.

3.3.2. Wastewater Engineering

In relation to the fish farm located in the central valley (municipality E), the students carried out field visits and meetings to collect information, in addition to a bibliographic review, from which they prepared a report that contains the characteristics of the production process, the main waste generated by the activity, a review of the existing information on flows, and physical–chemical and biological parameters. Based on the above points, they prepared proposals for improvements of the production process and the industrial waste treatment system.

In the case of the problem reported by municipality D, an analysis of the alternatives for the treatment of wastewater for rural areas was carried out. The above considered a report with a technical–economic proposal for the implementation of four alternatives for modular plants from various manufacturers of wastewater treatment systems.

3.3.3. Industrial Processes

Based on their interests, jobs, and activities expressed by the program members (including crafts, sewing, baking, and housework, among others), the students proposed creating three products using the most abundant solid waste in the community: plastic, glass, and used cooking oil. This initiative aimed to generate useful and/or decorative items with a high labor-added value, potentially serving as a source of income for the participants.

Residents of the coastal area were trained with recycling technique options via civil environmental engineering, such as making soap from domestic waste and reusing glass and plastics. Figure 5 shows images of the products obtained by the students.

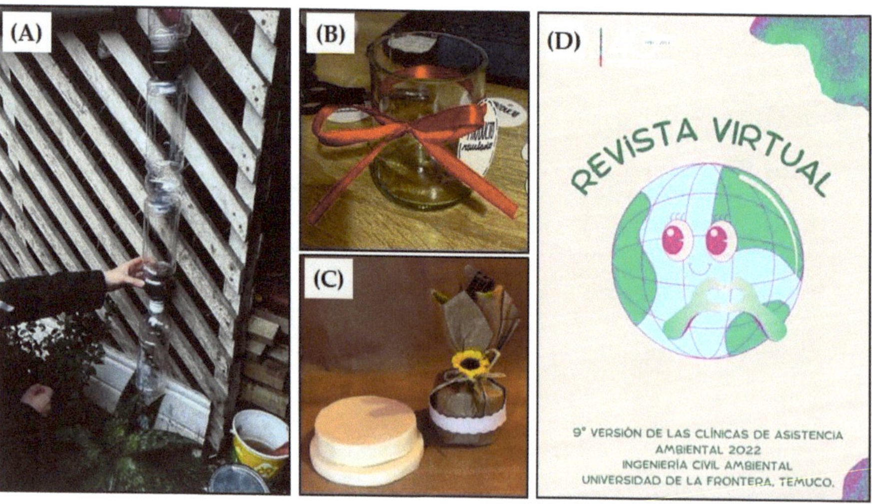

Figure 5. Products obtained by the students and presented to the members of the social programs: (**A**) a vertical garden, (**B**) glass, (**C**) soap, and (**D**) a digital magazine.

The execution of this training activity forced the students to study the different waste management options, in addition to looking for elements that would allow them to carry out practical activities to get to know their citizens better. Furthermore, implicitly, the students had to face a real audience that questioned and discussed each of the activities that were carried out.

3.3.4. Solid Waste Engineering

The students projected the population and generation of waste, analyzed different waste management possibilities (recycling, composting and anaerobic plants, and inorganic material recovery plants), and determined the minimum technical and economic requirements. The product obtained was a technical report with waste management alternatives for the city. This product allows municipal authorities to carry out a conceptual analysis with the purpose of projecting and determining the feasibility of designing and executing a large-scale project, prioritizing the scarce resources available from the municipality for these purposes.

3.3.5. Engineering Project

A composting plant for solid organic waste was designed, and the operation method, location options, and regulatory requirements of the projects were considered. The designed composting plant makes it possible to identify the ideal method of operation, the minimum necessary equipment and machinery, the critical infrastructure, and the costs associated with investment, operation, and maintenance, thus allowing for the dimensioning of what it means to have a municipal composting plant.

At the end of the semester, a ceremony was organized to deliver the products to the municipalities, in which each of the students made a presentation regarding the work they carried out, and also received a participation certificate from the MMA.

4. Discussion

Good technical performance by students was obtained; students learned by creating meaning from real day-to-day activities. In fact, the collaboration between the university and the municipalities offers a real work scenario that allows students to visualize the scenarios of their future work, similar to what was pointed out by Zhang [27] in the relationship between universities and companies. In this sense, knowledge acquisition was obtained by transferring the classroom into practice by incorporating experiences in students, generating opportunities for actively acquiring learning in a real context. On the other hand, in studies carried out by other authors on engineering students, another advantage was pointed out, which was related to the fact that engineering students increased their environmental awareness by taking subjects with educational content [28].

Therefore, it is essential to situate learning that implies locating thoughts and actions in a specific place and time. Situating means engaging other students, the environment, and activities to create meaning, finding in a particular environment the thought and action processes used by experts to perform knowledge and skilled tasks [29] (Figure 6). In the classroom, in this experience, it was possible to situate learning by creating conditions for participants to experience the complexity and ambiguity of learning in the real world. This is how the students who participated learned from their experiences, highlighting the relationships with the other participants, along with the activities, environmental signals, and social organization the community develops. In this sense, it is essential that students are committed to their assigned task [30]. The ECC approach has been positively valued by students [27] since it allows them to relate early on to real scenarios. At the end of the activity, the students indicated that they felt better prepared to enter the world of work, since they faced real problems, had to find solutions, and satisfy clients (municipalities). In fact, when students face real problems, they have the opportunity to devise solutions collaboratively, expanding their skills of analysis, creation, and expression of their ideas and networking.

The ECC requires them to develop social skills, generic skills (teamwork and leadership), and the creativity to solve real problems through analysis, reflection, and trial and error, where errors are seen as opportunities for improvement that stimulate them to fulfill their purposes. They leave the role of the student, and it allows them to become aware of what they are training for. Motivating them to identify the root cause of different problems in order to influence people's well-being, it was extremely satisfying for them to see how the solutions generated by them influence people's quality of life and realizing this stimulated them to move forward with great motivation and commitment [31,32].

On the other hand, the municipalities have expressed their agreement with the results obtained, since it allows them to obtain information for decision making in environmental matters. From the other point of view, the ECCs allow for sustainability, since the collaboration of the university in solving socio-environmental problems is necessary, considering the degree of specialization of PhDs. In this sense, it is important to keep in mind that for the success of the ECC it is necessary to have teachers with a high degree of specialization, since students need to be constantly guided in the search for solutions to the problems posed.

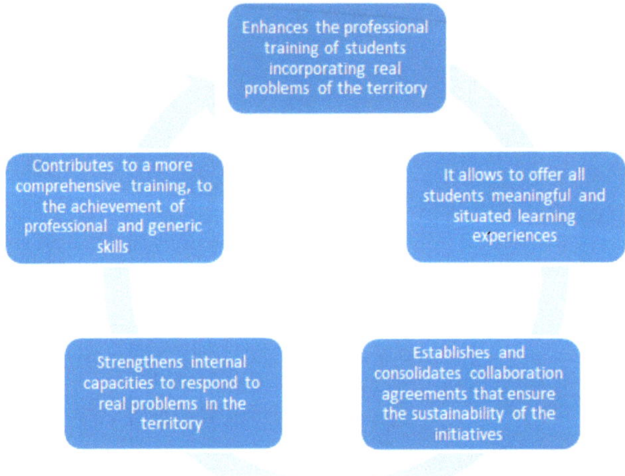

Figure 6. Situated learning essentially derives creative meaning from real problem solving.

From the point of view of managing the implementation of the ECC, it is important to consider that a person responsible for the UFRO and the MMA is permanently required for the implementation of the ECC; in addition, these professionals must have the technical capacity to analyze the various situations from a global perspective and also exercise leadership that allows symbiotically linking the university and the public institutions of the state.

At the conclusion of the first version of the ECC, there were requirements to incorporate other undergraduate courses from the Faculty of Engineering and Sciences to the ECC, which demonstrates the great interest in the execution of this practice; however, the ECCs were conceived to solve environmental problems, but this model is scalable to other types of problems. Considering that other undergraduate programs may have other learning objects, it is always possible to apply the methodology, changing the type of problems to be solved.

The ECCs have managed to position themselves among the students and municipalities of the region; this methodology gave rise to the licensing contract being signed in January 2022 between the Ministry of the Environment and the University of La Frontera. Thus, what started as an undergraduate activity has now become a protocol that protects itself; it has become a license for a technology transfer from the university to the environment, the most concrete way to transfer knowledge and technology.

It should be noted that this innovative educational experience also gave rise to the creation of the Node for Strengthening Local Environmental Governance (Spanish acronym "NOGAL") of the Ministry of the Environment of Chile. In this way, knowledge and technology will be transferred through a license. This experience has been licensed to other universities, such as the Catholic University of the North, Austral University of Chile, the

Puerto Montt branch, and six different universities, which demonstrates how innovative this methodology is and the advantages involved in implementing it with undergraduate students since it is active learning located in a real context that also collaborates in the solution of local and real problems, which do not have resources for solution management.

5. Conclusions

The results show that the correct implementation of the ECC active learning methodological proposal depends on a series of factors, such as the professional capacity of the designated counterparts from the MMA and the university, the adequacy of the teachers to employ the new method, and the ability to respond to the requirements of the municipalities.

After this first experience of implementing the ECC, new versions have been developed, all with a high percentage of subject participation, incorporating other new undergraduate studies, and a high demand for participation from municipalities, which demonstrates that this methodology has been consolidated in its application.

The subjects that presented the greatest number of requirements corresponded to environmental analytical chemistry and wastewater engineering, in the first case due to the need to know the baseline of water bodies and, in the second case, the search for solutions for the treatment of solid and liquid wastes, which generates great impacts on the environment.

The development of activities in a real context allowed for the classroom to be transferred into a different scenario, which allowed for active learning to be experienced, developing students' skills, and acquiring knowledge experientially, which will enable us to train professionals who meet the needs of our society [27].

The implementation of the ECC has been positively valued by students since it allows them to face a real scenario, identifying early on how they will perform at work in the future. From the point of view of active learning located in real contexts, this methodology allows students to face real problems, where there are clients that they must satisfy, and to do so, they must be able to diagnose, plan, learn, and evaluate. This is consistent with the growing interest on the part of international agencies in higher education institutions in developing technical skills but also soft skills in real-life scenarios [33].

The PDCA cycle methodology allowed us to guide the students' work to satisfy municipal requirements, such as rainwater harvesting systems, composting plant projects, biodiesel plant projects from domestic waste oil, and environmental education strategies.

The methodological framework that has been introduced in this study seeks to serve as a guide for the universities of the world, which aims to develop active learning in its students, considering real contexts and a real connection with society in the search for solutions to the environmental problems present in the territories.

The main limitation of the implementation of the ECC methodology is that it requires a permanent administrative structure at the university and in the MMA, to grant autonomy, in addition to financial resources to carry out a number of field and end-of-semester activities. In this sense, the possibility of incorporating the implementation of the ECC into the academic careers of teachers, in addition to the incorporation of other degrees into the ECC, arises as an opportunity for improvement with the aim of solving problems from a more general perspective. Both improvement proposals provide autonomy, scope, and a comprehensive approach in the search for solutions to environmental problems.

Author Contributions: Conceptualization, M.Z.R. and D.R.-L.; investigation, M.Z.R., D.R.-L. and O.R.; methodology, M.Z.R. and D.R.-L.; writing original draft, M.Z.R., J.V. and D.R.-L.; reviewing and editing, M.Z.R., D.R.-L., M.A. and O.R. All authors have read and agreed to the published version of the manuscript.

Funding: This research was funded by ANID/FONDAP/15130015 and ANID/FONDAP/1523A0001 projects and Faculty of Engineering and Sciences of the Universidad de la Frontera.

Institutional Review Board Statement: Not applicable.

Informed Consent Statement: Not applicable.

Data Availability Statement: Data are contained within the article and Appendix A.

Conflicts of Interest: The authors declare no conflicts of interest.

Appendix A

Table A1. Application form.

Application Form Environmental Engineering Clinics		
1. General data		
Municipality		SCAM level:
Designated counterparty	Name	
	Position:	
	Telephone:	
	Email:	
Date		
2. Requirement data (mark only 1 subject per requirement)		
Select the subjects	x	Subject 1
		Subject 2
		. . . .
		Subject N
Description of the problem	Explain the problem in a maximum of 100 words.	
Result expected by the municipality	Explain in a maximum of 30 words the result that the municipality expects.	
Contribution of the municipality	Example: only incorporate pecuniary contributions. By en.: the municipality will provide transportation for students from the UFRO to the water sampling point.	
Does it agree with any line of the environmental strategy?	Yes / No Which?	

References

1. United Nations. *Sustainability Development Goals: 17 Goals to Transform Our World*; United Nations: New York, NY, USA, 2021. Available online: https://www.un.org/sustainabledevelopment/ (accessed on 20 June 2023).
2. Prieto-Saborit, J.A.; Méndez-Alonso, D.; Fernández-Viciana, A.; Dixit, L.J.D.; Nistal-Hernández, P. Implementation of Cooperative Learning and Its Relationship with Prior Training of Teachers, Performance and Equity in Mathematics: A Longitudinal Study. *Sustainability* **2022**, *14*, 16243. [CrossRef]
3. Silva, J.; Maturana, D. Una propuesta de modelo para introducir metodologías activas en educación superior. *Innov. Educ.* **2017**, *17*, 117–131.
4. Lima, R.; Andersson, P.; Saalman, E. Active Learning in Engineering Education: A (re)introduction. *Eur. J. Eng. Educ.* **2017**, *42*, 1–4. [CrossRef]
5. Martinez, E.; Pegalajar, M.C.; Burgos-Garcia, A. Active methodologies and curricular sustainability in teacher training. *Int. J. Sustain. High. Educ.* **2023**, *24*, 1364–1380. [CrossRef]
6. Coll, C.; Mauri, T.; Onrubia, J. Análisis y resolución de casos-problema mediante el aprendizaje colaborativo. *Rev. Univ. Soc. Conoc.* **2006**, *3*, 29. [CrossRef]
7. Mintz, K.; Tal, T. The place of content and pedagogy in shaping sustainability learning outcomes in higher education. *Environ. Educ. Res.* **2016**, *24*, 207–229. [CrossRef]
8. UNESCO. *Las Tecnologías de la Información y la Comunicación en la Formación Docente*; Informe UNESCO: Paris, France, 2004.
9. Milovanovic, J.; Shealy, T.; Katz, A. Higher Perceived Design Thinking Traits and Active Learning in Design Courses Motivate Engineering Students to Tackle Energy Sustainability in Their Careers. *Sustainability* **2021**, *13*, 12570. [CrossRef]

10. Atienza, J. *Aprendizaje Basado en Problemas*; En Labrador, M., Andreau, M., Eds.; Metodologías Activas; Ediciones Universidad Politécnica de Valencia: Valencia, Spain, 2008; pp. 11–24.
11. Schneider, E.; Froze, I.; Rolon, V.; Mara de Almeida, C. Sala de Aula Invertida em EAD: Uma proposta de Blended Learning. *Rev. Intersaberes* **2013**, *8*, 68–81. [CrossRef]
12. Michaelsen, K.; Davidson, N.; Major, C. Team Based Learning Practices and Principles in Comparison with Cooperative Learning and Problem Based Learning. *J. Excell. Coll. Teach.* **2014**, *25*, 57–84.
13. Lopes, J.; Viegas, M.C.; Pinto, A.J. (Eds.) The Importance of Making Teaching Practices Public, Shareable, and Usable. In *Multimodal Narratives in Research and Teaching Practices*; IGI Global: Hershey, PA, USA, 2019; pp. 1–43.
14. Clancey, W.J. A tutorial on situated learning. In *Emerging Computer Technologies in Education: Selected Papers of the International Conference on Computers and Education (Taiwan)*; Self, J., Ed.; AACE: Charlottesville, VA, USA, 1995; pp. 49–70.
15. Stein, D. Situated Learning in Adult Education. 1998. Available online: https://www.edpsycinteractive.org/files/sitadlted.html (accessed on 14 October 2023).
16. Vygotsky, L.S. *Mind in Society: The Development of Higher Psychological Processes*; Cole, M., JohnSteiner, V., Scribner, S., Souberman, E., Eds.; Harvard University Press: Cambridge, MA, USA, 1978.
17. Vygotsky, L.S. *Thought and Language*; Kozulin, A., Ed.; MIT Press: Cambridge, MA, USA, 1986.
18. Paineán, B.; Aliaga, O.; Torres, T. Aprendizaje basado en problemas: Evaluación de una propuesta curricular para la formación inicial docente. *Estud. Pedagóg.* **2012**, *38*, 161–180. [CrossRef]
19. Mahnashi, I.; Salah, B.; Ragab, A.E. Industry 4.0 Framework Based on Organizational Diagnostics and Plan–Do–Check–Act Cycle for the Saudi Arabian Cement Sector. *Sustainability* **2023**, *15*, 11261. [CrossRef]
20. Prashar, A. Adopting PDCA (Plan-Do-Check-Act) cycle for energy optimization in energy-intensive SMEs. *J. Clean. Prod.* **2017**, *145*, 277–293. [CrossRef]
21. Zambrano, M.; Rodríguez, D. Manual Clínicas de Asistencia Ambiental: Aprendizaje en Contexto Real, Para la Formación Integral de Ingenieros. Registro de Propiedad Intelectual No 297.453. 2018. Available online: https://www.propiedadintelectual.gob.cl/ (accessed on 2 November 2023).
22. National Statistics Institute. *Synthesis of Results: Census 2017*; Government of Chile: Santiago, Chile, 2018. Available online: https://www.ine.cl/docs/default-source/censo-de-poblacion-y-vivienda/publicaciones-y-anuarios/2017/publicaci%C3%B3n-de-resultados/sintesis-de-resultados-censo2017.pdf?sfvrsn=1b2dfb06_6 (accessed on 2 November 2023).
23. Rodríguez-Luna, D.; Vela, N.; Alcalá, F.J.; Encina-Montoya, F. The Environmental Impact Assessment in Aquaculture Projects in Chile: A Retrospective and Prospective Review Considering Cultural Aspects. *Sustainability* **2021**, *13*, 9006. [CrossRef]
24. Deming, W.E. Total Quality Management in Higher Education. *Manag. Serv.* **1993**, *35*, 18–20.
25. Jones, E.C.; Parast, M.M.; Adams, S.G. A framework for effective Six Sigma implementation. *Total Qual. Manag. Bus. Excell.* **2010**, *21*, 415–424. [CrossRef]
26. Chilean Regulation 1333 Water quality requirements for different uses. In *Official Journal of Chile*; Government of Chile: Santiago, Chile, 1978.
27. Zhang, Y.; Chen, X. Empirical Analysis of University–Industry Collaboration in Postgraduate Education: A Case Study of Chinese Universities of Applied Sciences. *Sustainability* **2023**, *15*, 6252. [CrossRef]
28. Nogueira, T.; Castro, R.; Magano, J. Educación de estudiantes de ingeniería en sostenibilidad: El papel moderador de la inteligencia emocional. *Sustainability* **2023**, *15*, 5389. [CrossRef]
29. Lave, J.; Wenger, E. *Situated Learning*; Cambridge University Press: New York, NY, USA, 1991.
30. Peng, L.; Deng, Y.; Jin, S. The Evaluation of Active Learning Classrooms: Impact of Spatial Factors on Students' Learning Experience and Learning Engagement. *Sustainability* **2022**, *14*, 4839. [CrossRef]
31. Emblen-Perry, K. Auditing a case study: Enhancing case-based learning in education for sustainability. *J. Clean. Prod.* **2022**, *381*, 134994. [CrossRef]
32. Neves, R.M.; Lima, R.M.; Mesquita, D. Teacher Competences for Active Learning in Engineering Education. *Sustainability* **2021**, *13*, 9231. [CrossRef]
33. Ariza, J.Á.; Olatunde-Aiyedun, T.G. Bringing Project-Based Learning into Renewable and Sustainable Energy Education: A Case Study on the Development of the Electric Vehicle EOLO. *Sustainability* **2023**, *15*, 10275. [CrossRef]

 *sustainability*

Article

Instructors' Perspectives on Enhancing Sustainability's Diffusion into Mechanical Engineering Courses

Joan K. Tisdale [1],* and Angela R. Bielefeldt [2]

[1] Integrated Design Engineering Program, University of Colorado Boulder, 522 UCB, Boulder, CO 80309, USA
[2] Department of Civil, Environmental & Architectural Engineering, University of Colorado Boulder, 428 UCB, Boulder, CO 80309, USA; angela.bielefeldt@colorado.edu
* Correspondence: joan.tisdale@colorado.edu; Tel.: +1-303-735-4426

Abstract: This research strives to catalyze a more extensive integration of sustainability topics into mechanical engineering (ME) courses. The process through which higher education instructors choose to integrate sustainability topics into their courses was conceptualized using diffusion of innovation theory. The research explored two questions: (1) What factors were influential to front runners (innovators or early adopters) for sustainability integration in undergraduate courses taken by ME students? (2) What factors could spur non-adopters to integrate sustainability into their courses? The study included a survey (with 53 respondents who taught sustainability and 14 respondents who did not teach sustainability), 10 interviews with innovators and early adopters, and a focus group of 5 participants. The results were explored primarily from the perspective of meeting the needs of instructors (the target users). Peer-to-peer interactions were found to be important across all user groups. Practices that would help motivate later adopters include prepared curriculum- and university- or department-based support in the form of mission statements, training, mandatory sustainability inclusion, and a sustainability office to provide support. Diffusion of innovation theory provides insights into which strategies are likely to be most effective in expanding the number of faculty members who integrate sustainability topics into their courses.

Keywords: sustainability education; engineering education; higher education; diffusion of innovations; propagation; faculty innovators

check for updates

Citation: Tisdale, J.K.; Bielefeldt, A.R. Instructors' Perspectives on Enhancing Sustainability's Diffusion into Mechanical Engineering Courses. *Sustainability* **2024**, *16*, 53. https://doi.org/10.3390/su16010053

Academic Editors: Clara Viegas and Natércia Lima

Received: 1 November 2023
Revised: 14 December 2023
Accepted: 18 December 2023
Published: 20 December 2023

1. Introduction

Engineers have an important role to play in contributing to sustainable development. This requires educating engineers about sustainability in order to prepare them for this challenging endeavor. While the literature contains many examples of how instructors in higher education are teaching sustainability [1], the United Nations Sustainable Development Goals (SDGs) [2], and sustainability in the context of engineering [3–5], more widespread integration would be beneficial [6]. For example, a recent study found that 57 of 100 programs showed no evidence of integrating sustainability topics into any of the required ME courses in their curricula [7].

The process through which instructors decide what to teach and how to change their courses over time has been conceptualized via a variety of frameworks and theories. The Academic Plan model of Lattuca and Stark [8] describes the iterative process of curriculum decisions as being affected by internal influences (e.g., personal, unit-level, and institutional), external influences (e.g., market forces and accreditation requirements), the educational environment and outcomes, and the broader sociocultural context. This model has been previously used to explore faculty decisions related to teaching macroethics in engineering [9] and diversity issues in environmental and sustainability programs [10]. Education for sustainability enhancement at a program level has been explored through change management perspectives [11] and the Four Categories of Change Strategies model of Borrego and Henderson [12]. Diffusion of innovations (DoI) theory has been previously applied to study changes in higher education [13–16]. In higher education settings, the unit of analysis for DoI could be individual instructors, departments, or institutions [17]. No studies applying DoI to sustainability teaching in higher education were found. This research elected to use the DoI model in the hope of providing new insights with respect to individual instructor decisions to incorporate sustainability into their ME courses.

Diffusion of innovations (DoI) theory seeks to explain how innovations are adopted in a population, and an innovation can be an idea, behavior, or object that is perceived as new by its audience [18–20]. DoI concepts explain how an innovation gains momentum and diffuses or spreads through a specific population or social system over time. The result of this diffusion is that people, as pieces of a larger social system, adopt the new idea, behavior, or product. Important factors that impact the rate at which innovations are adopted include the attributes of the innovation itself (e.g., relative advantage and compatibility), the types of innovation decisions (e.g., optional vs. mandatory), communication channels, the nature of the social system, and the extent of change agents' promotion efforts [13,18,21].

DoI theory classifies a population into five different segments based on their propensity to adopt an innovation [18,19]. The first individuals to adopt an innovation are termed innovators, and they make up around 3% of the market. They are followed by early adopters and the early majority, the next ~13% and 34% of individuals who adopt the innovation, respectively. After approximately half of the market has adopted an innovation, those in the late majority (the next 34%) and the laggards (the final 16%) adopt an innovation that achieves full penetration. Research has found that users in these different population segments have unique needs and characteristics [18–20]. Thus, strategies to promote an innovation should adapt and change over time, depending on the extent of market penetration that has been achieved.

It is difficult to gauge the extent of sustainability integration by faculty members in mechanical engineering (ME). A previous study found that 45% of 151 ME instructors taught sustainability [22], but it was assumed that survey respondents were over-represented among those who teach ethics and societal impact topics due to leverage salience theory. In a benchmarking study of ME programs at 100 institutions, 17 of the ME programs offered no undergraduate courses that integrated sustainability, while 9 programs had 13 to 38 undergraduate courses with sustainability, based on information submitted to the American Association for the Advancement of Sustainability in Higher Education (AASHE) and university course catalogs [1]. Within these programs, it is unclear what percentage of the ME instructors have integrated sustainability into their courses, but it seems that, depending on their institution, an individual could be a local innovator or part of the late majority. In addition to individual instructor attributes, the courses they were assigned to teach seemed relevant because particular courses, like thermodynamics and design, more commonly integrated sustainability than others (e.g., statics).

Mechanical engineering was selected as the discipline of interest in this study due to its high impact. Engineering has significant environmental, social, and economic impacts and, thus, plays an important role in global sustainable development efforts. The largest number of engineering bachelor's degrees in the U.S. are awarded in ME [23], so it is crucial that mechanical engineers entering the workforce have the necessary knowledge, skills, and attitudes to promote sustainable practices. There are many published examples of sustainability's integration into ME courses (e.g., Enelund et al. [24], Ramanujuan et al. [25], Issa [26], and Jha [27]). However, ME has been found to be lagging behind other engineering disciplines (e.g., environmental, civil, chemical, and material) in teaching sustainability [22,28–31].

The DoI framework can go hand in hand with the Academic Plan model, and it was selected for its unique perspectives on various user groups. Thus, this study used DoI as a lens to explore factors that contribute to instructors' integrating sustainability topics into undergraduate courses for mechanical engineering students. The specific research questions were as follows:

- What factors were influential to instructors who were self-described front runners (innovators or early adopters) in sustainability's integration into undergraduate courses taken by mechanical engineering students?
- What factors could spur non-adopters to integrate sustainability into their higher education courses?

2. Materials and Methods

The research used a mixed-methods approach, starting with a survey followed by interviews and a focus group.

2.1. Survey

A survey was developed to identify basic information about the sustainability teaching motivations and practices of instructors of ME undergraduate courses. The primary goal of the survey was to support the criterion-based selection of individuals to recruit for interviews. The survey was developed based on the research questions and information available in the literature. The survey first asked individuals to identify whether or not they taught undergraduate courses for ME students that included sustainability. Those who answered affirmatively were then asked about the following: what inspired them to incorporate sustainability in their undergraduate courses (6 options provided + 1 other, fill-in); what was helpful when they were first integrating sustainability (3 options + 1 other, fill-in); the courses they taught that included sustainability (7 options + 2 others, fill-in); and the sustainability-related topics they taught (5 options + 1 other, fill-in). For those who indicated that they did not teach sustainability in their undergraduate engineering courses, the follow-up questions concerned the reasons sustainability had not been included (5 options + 2 others, fill-in) and what would encourage them to incorporate sustainability (6 options + 1 other, fill-in). All surveys also asked respondents a final open-response question inviting them to share anything else about teaching sustainability in mechanical engineering and whether they wanted to volunteer to participate in a focus group. More details of the survey were provided by Tisdale [31].

Individuals believed to represent potential front runners of sustainability education in ME were invited to participate in the survey. These individuals were identified using the following criteria: (1) those who had authored papers on sustainability education in ME, thermodynamics courses or engineering design courses; (2) those who had received a grant from the U.S. National Science Foundation for sustainability integration into ME [32]; (3) instructors teaching an ME course that integrated sustainability based on documentation submitted by their institutions to AASHE toward certification under the Sustainability Tracking, Assessment & Rating System (STARS) (see Tisdale and Bielefeldt [1] for more information); (4) all instructors within ME programs at the "top 10" institutions (8 inside the U.S. and 2 outside the U.S.) based on sustainability's inclusions in the largest number of undergraduate ME courses [1]; (5) respondents to a 2016 survey who checked yes to teaching both sustainability and ME students [22]. University websites were used to retrieve email addresses. Instructors identified in each of the above groups were contacted via email with a survey invitation and a link to the online survey in Qualtrics; 564 invitations to instructors at 125 different institutions were successfully delivered. The email also invited individuals to forward the survey invitation to other faculty members whom they knew taught sustainability to ME undergraduate students to foster snowball sampling. Survey responses were gathered in January to February 2023.

Among the survey responses that included useful information, 53 respondents indicated that they taught sustainability in courses for undergraduate engineering students (representing 38 different institutions, including 4 outside the U.S.). The most common courses where the survey respondents had integrated sustainability were senior capstone ($n = 25$), renewable/sustainable energy ($n = 22$), first-year design ($n = 13$), and thermodynamics ($n = 13$) courses. The most common sustainability topics taught included environmental issues, followed by economic issues, social issues, renewable energy, and global development/SDGs. There were also 14 survey respondents who did not teach sustainability in undergraduate courses (representing 9 different institutions, including 6 institutions with respondents who taught sustainability).

2.2. Interviews

Interviews were used to learn from instructors in the innovator and early adopter groups about their experiences with sustainability's inclusion in ME courses and programs. There were 24 survey respondents who volunteered to participate in interviews. Fifteen individuals who taught sustainability in a wide range of ME courses and institutional settings were invited to participate in interviews, and 10 interviews were completed.

Semi-structured interviews were conducted over Zoom for about 30–60 min and recorded. The interviewees were asked questions such as where they placed themselves in the DoI model adopter categories in terms of integrating sustainability into mechanical engineering courses and why, whether they believed their sustainability teaching practices had influenced others, and whether they had suggestions for encouraging other faculty members to integrate sustainability in their courses and/or curricula. A full list of the interview questions was provided by Tisdale [31].

The interviewees taught and/or previously taught at both public ($n = 5$) and private institutions ($n = 8$) (three individuals described two different institutions), institutions from a range of Carnegie classifications (six with very high doctoral research activity, two with high doctoral research activity, one doctoral/professional university, two master's universities, one special-focus four-year university, and one large public research-intensive institution outside the U.S.) with a variety of locations and institution sizes (small, medium, and large, based on the number of undergraduate students). The interviewees each had 5 to over 21 years of university teaching experience, were tenured/tenure-track ($n = 5$), and were teaching faculty members/instructors ($n = 5$). They had degrees from a range of different disciplines (mechanical engineering, $n = 8$, other engineering disciplines, $n = 5$, and non-engineering, $n = 2$), and they were male ($n = 4$), female ($n = 5$), and non-binary ($n = 1$). Based on their survey responses, the interviewees taught sustainability in courses that were related to renewable/sustainable energy ($n = 7$), senior capstone courses ($n = 5$), first-year design ($n = 5$), thermodynamics ($n = 2$), heat transfer ($n = 2$), manufacturing ($n = 2$), and a variety of other required and elective courses ($n = 8$); the most commonly integrated topics were environmental, economic, social, renewable/sustainable energy, global development/SDGs, and others (e.g., concerning optimization, circular economy, life cycle assessment, metrics, ethical implications, environmental justice, food systems, nature-inspired, and collapse). To maintain anonymity, the interviewees are referred to with numbers, e.g., Interviewee 1.

2.3. Focus Group

A 1-h focus group over Zoom was conducted with 5 participants (3 individuals who were interviewed and 2 new individuals). The topics included the following: whether they had witnessed any specific resources or sustainability inclusion types that sparked greater interest and inclusion; key factors that led to a sustainability integration that withstood the test of time; the importance (or lack thereof) of a cohort or collaboration among faculty members; effective examples of top-down and bottom-up approaches; and the role of the institutional and educational environment in sustainability adoption.

2.4. Analysis: Coding

Zoom created a transcript of the audio from the recordings of the interviews, and these were then edited manually to correct errors. Deductive and inductive coding were utilized to systematically analyze the interview and focus group transcripts [33]. The two overarching themes of peer-to-peer communication and user-based needs from DoI theory formed the foundation of the coding scheme. Subtypes within these themes were identified, and a constant comparative method was used across the transcripts. The two authors discussed the themes to arrive at an agreement. The codes are shown in Table 1. For their presentation in this paper, clean verbatim quotes are used from which filler language, such as "like" and "um", were removed.

Table 1. Influences on faculty members' integration of sustainability into their teaching: codes, definitions, and number of responses.

Broad Category	Subcategory	Definitions	Interview *n* of 10	Survey, *n*	
				Teach Sustainability, of 53	Do Not, of 14
Self-realization of importance		Personal realization of importance due to life experiences, including global travel, becoming a parent or grandparent, and others	5	45 ^	3
Peer-to-peer	Influenced their adoption	Peer-peer conversations, networks, and community; direct interactions, oftentimes with others who share common interests and/or goals, which lead to the adoption of an innovation	4	1 *	NA
	They influenced others		9	0	NA
	Publications	Reading a paper or textbook	3	7 *	0
Faculty (user-based) needs					
University-based	Mission and policy	Department, college, or university mission, priorities, policy, and/or support	4	17	7
	Incentives	Reasons and motivations to take the steps and make the sacrifices needed	1	3 *	2 +
	Mentor program	A one-to-one relationship in which a more experienced faculty member can share wisdom, insight, and tools with the goal to assist the less-experienced member	1	13	1
	Sustainability office	An office focused on sustainability support for the university, including assisting professors in their sustainability-related goals	2	1 *	0
Resources	Prepared curriculum	Teaching materials that are ready for instructors/professors to utilize in the course	3	14	7
	Training	Offer instruction for the acquisition of the desired knowledge and expertise	3	7	5
Organization	Bandwidth	The time, openness, and place for the topic to fit in the course or for the course to fit into the major's four-year plan	1	0	8
	Consistency	A standard set of definitions and a curriculum associated with the term "sustainability"	1	0	1
Meeting student needs	Interest	Engaging, intriguing, desirable, and relatable to students	10	5 *	0
	Whole-person	The ability to be a whole person with emotions Clear opportunities for hope and empowerment	3	0	0

^ Survey option: a personal passion; a number of the "other" responses also elaborated on this; + survey option: a bonus or financial incentive; * coded write-in responses.

2.5. Limitations

This work was limited by the number and experience of the interviewees, focus group participants, and survey respondents. Increasing these numbers would have made the study more comprehensive. Among the 14 survey respondents who did not teach sustainability in ME, 11 were ME faculty members in 6 programs with atypically strong integrations of sustainability in ME. The sources for contacts were U.S.- and English-centric and thus, the experiences of other faculty were largely absent. In addition, the clear classification of instructors into user groups was uncertain, based on local norms and facets of sustainability.

3. Results and Discussion

The results include a description of how the interviewees classified themselves in the DoI user groups and the key influences found for the adoption of sustainability education in engineering.

3.1. Adopter Characterization

Based on the process used to select those who were interviewed, the authors classified the interview participants as innovators or early adopters of sustainability's incorporation into undergraduate ME education. Four interviewees categorized themselves as innovators (Interviewees 6, 7, 8, and 10), five as early adopters (Interviewees 2, 3, 4, 5, and 9), and two as an early majority (Interviewees 1 and 4), with one interviewee citing two adjacent categories and other interviewees noting specific ways in which they were particularly innovative. Interviewee 10 characterized herself as a risk-taker and one of the first in engineering but noted that some colleagues on campus had been working on sustainability since the 1970s and that engineering "is quite late in all of these discussions". Further, she stated that she had "senior colleagues who have always worked in this space who are quite quiet... perhaps they got a lot of pushback in the beginning... have decided to face outward internationally to colleagues who are like minded..." Another innovator, Interviewee 6, noted that "one makes mistakes because you are the first in". Looking back, they "approached as an engineer", which had limitations, and they "made enemies of colleagues and was perceived as a zealot". Interviewee 4 felt like an early adopter of social sustainability. Most of the interviewees indicated uncertainty concerning what others were doing at their own institutions, nationally, and/or globally. Interviewee 3 stated that "all [mechanical] educators [teaching sustainability] are still early adopters". Thus, the survey respondents who were teaching sustainability might all be considered innovators or early adopters, while the survey respondents who were not teaching sustainability might become an early majority (e.g., interested enough in sustainability to respond to the survey but not yet teaching sustainability in their undergraduate courses).

The results indicate challenges with characterizing an innovation itself—such as any integration of sustainability into engineering and/or mechanical engineering, teaching sustainability in particular courses (e.g., design or statics), or teaching specific concepts (e.g., environmental life cycle assessment or social impacts). In the survey, two write-in responses indicated the respondents might teach sustainability if they were assigned to particular courses (such as thermodynamics) but that sustainability did not "fit" into their courses (dynamics and automatic controls). In contrast, one survey respondent stated, "Frankly sustainability should be integrated into the whole mechanical engineering curriculum. In my view there is not one traditional mechanical engineering topic that in some way could not address sustainability". Furthermore, local context mattered. For example, Interviewee 8 noted that nearly everyone in his small program had probably taught sustainability due to courses like first-year projects and senior design in which there was a lot of team or collaborative teaching; in contrast, others worried about whether anyone in their department would continue to teach sustainability after they had retired (e.g., Interviewee 5).

3.2. Personal Life Experience

Among the innovators and early adopters, there were not necessarily models or incentives to teach about sustainability. Rather, many of the interviewees described personal experiences in their lives that led them to realize the critical importance of sustainability and decide to integrate these topics into their teaching. Five interviewees described seeing and experiencing pollution and poverty conditions globally during their travels as influential. Three interviewees described personal motivations concerning the importance for future generations associated with becoming parents or grandparents. Within the survey, the most commonly selected response to the question "what inspired you to incorporate sustainability in your undergraduate courses" was a personal passion (n = 45 of 53); this passion had perhaps originated via personal life experiences. One of the write-in responses noted a "Deep concern about the climate crisis. This is why I went into engineering". This motivator is likely to grow as global impacts become more readily apparent.

3.3. Peer-to-Peer Communication

Peer-to-peer communication has been identified in DoI theory as an important aspect of the adoption of new behaviors, particularly when moving from early adopters to majority audiences [19]. Peer-to-peer interaction was important in their sustainability teaching practices for four interviewees, and nine interviewees also described how their direct interactions had led others to integrate sustainability into their teaching. Mediums for peer-to-peer communication could include the following: general conversations with friends, colleagues, and peers; the inner- and inter-departmental sharing of ideas and materials; guest speaking and conferences; and the general sharing of ideas among groups sharing their goals and motivations.

Three interviewees articulated how sustainability's incorporation had spread to them and how they had then continued this diffusion. Interviewee 9 explained, "I saw a presentation at ASEE [the American Society for Engineering Education] about 10 years ago, by someone who was incorporating sustainability in their classroom... I went up to him and asked, could I get some of those resources? And he shared his entire drop-box with me". Interviewee 9 later shared her work with peers at her institution and via conference presentations. This was a powerful example of peer-to-peer impacts on the diffusion of sustainability both to and from the interviewee.

Four interviewees described peer-to-peer examples of their roles in diffusing sustainability education in their own colleges and universities. Interviewee 2 and Interviewee 9 described how they had created sustainability-inclusive course materials in senior design and first-year introduction to engineering courses, respectively, which were then adopted by other faculty members at their institution, thus diffusing the innovation of sustainability inclusion. Interviewee 3 indicated that professors in other departments at his institution reached out to him as his reputation for sustainability work grew. Interviewee 7 described that they were "frequently invited to speak in other departments" about their sustainability teaching practices, and they also discussed how their sustainability work with students that was supported by National Science Foundation grants spread both nationally and internationally.

Publications appear to be important to DoI in academic settings. Publications might be viewed as a form of asynchronous peer interaction, similar to the example in which two women engineering faculty members identified books as sources of mentoring [34]. Interviewee 7 stated, "I wrote a paper in 2008 which now has about 15,000 reads on Research Gate". Eight of the 10 interviewees had published papers describing their sustainability teaching in engineering courses, but only three discussed these in their interviews. In the survey, write-in comments from five respondents indicated that they found reading to be helpful when they were first integrating sustainability into a curriculum, and another two indicated a textbook; another four cited research, which might entail researching the topic through reading the literature. With publications, faculty members are educating other professors on ways to effectively integrate sustainability into several types of courses.

3.4. User Needs: Instructors

Instructors are the users of this innovation because they decide whether or not to teach sustainability in mechanical engineering courses. The interviewees described their own needs (both past and/or present), as well as their understanding of the needs of those in the later user groups, in the adoption of sustainability education. The survey respondents and focus group participants were in varied user groups, from innovators in sustainability integration to those lacking sustainability integrations. Based on the interviews, areas of professor-based user needs were categorized in the sub-topic areas described in Table 1, including university-based support, resources, organization, and meeting student needs.

3.4.1. University-Based Support

University-based support represents general and/or financial support from the dean, chancellor, or others at a university. University support can come in multiple forms, including supportive policy from the college and/or university, incentives, mentor opportunities, and partnerships with a sustainability office. For example, a focus group participant stated, "I feel like there needs to be something above the individual faculty level, like department level or college level that's encouraging it, and I've noticed that's been successful on our campus".

Policies supporting sustainability education from a university can be very impactful. There were 15 of the 53 survey respondents with sustainability in their undergraduate engineering courses who indicated that "a department, college or university mission inspired them to incorporate sustainability into their courses". Interviewee 1 explained that, at her university, a policy mandates that all undergraduate students are exposed to sustainability principles in one general-education course and one major-specific course. Therefore, all academic departments integrate sustainability in at least one required course. In this case, sustainability was primarily integrated into existing core courses, which did not add any courses to students' four-year curriculum. Further, Interviewee 8 worked at an engineering college that has a sustainability-dedicated course that is mandatory for all undergraduate engineering students. He described the benefit of this dedicated course as the ability to deeply explore sustainability topics and issues in creative and interactive ways. Interviewees 1 and 8 expressed confidence in their graduates having an awareness of and skills in sustainability. It appears that an administrative structure for sustainability requirements in curricula has been effective both in increasing sustainability integration in undergraduate mechanical engineering courses and in achieving sustainability learning outcomes among students. In considering whether sustainability should be included, Interviewee 5 stated, "This isn't just a nice course to have. This is the future of the world we're talking about, and so we need to make it a priority in our programs to incorporate this".

Incentives can play an important role in sustainability adoption. A focus group participant described this need: "You need some incentive to want to put the time in, to alter your questions and add a sustainability piece into whatever you're already doing". This was also expressed by Interviewee 1, who said, "So if you've got mentors to help figure out how to integrate it into a course or monetary support to change a course, that's going to motivate different people to do it. I think because we're all so busy in what we do, having some kind of incentive to incorporate it more into your curriculum is needed, but also having the support of your department chair, your dean, or the Chancellor...all those pieces are necessary to really propagate it". In responding to the survey question regarding what would encourage them to incorporate sustainability into one course or more of their courses, the responses among the 14 respondents who were not currently teaching sustainability were as follows: seven cited department incentives, three cited a bonus or financial incentives, and two cited sustainability's inclusion as part of their performance reviews.

A partnership with a sustainability office on campus is another support for integrating sustainability into courses. One survey respondent stated that the sustainability office at their university was very helpful to them when they were first integrating sustainability into the curriculum. AASHE STARS 2.2 also awards a point if a campus offers sustainability coordination via a committee, office, or officer [35]. A successful example of this is the Green Academy at Nottingham Trent University, which reaches out to all schools and departments at the university to help with the incorporation of sustainable development in their curricula [36].

Among the 53 survey respondents who taught sustainability in their undergraduate courses, 13 indicated that **mentorship** was helpful to them when they were first incorporating sustainability. In addition, one of the individuals who was not teaching sustainability indicated on the survey that a mentor would encourage them to incorporate sustainability into one or more of their courses. Mentoring has been found to foster collegiality and collaboration while increasing teaching and research productivity [37]. Mentoring contributes to building positive relationships, and it can encourage professors to challenge assumptions underlying current unsustainable practices [38].

3.4.2. Resources

A number of faculty members indicated that they did not feel prepared to integrate sustainability topics into their courses, and outside resources might help these faculty overcome this obstacle. Five of the 14 survey respondents who did not teach sustainability in their mechanical engineering courses cited their lack of expertise in sustainability as a reason. Interviewee 4 echoed this sentiment: "I think there are still some people who are afraid to approach some of these topics". A prepared curriculum and/or training could help meet these needs.

A prepared curriculum was noted as helpful in both the interviews and the survey. Six of the 14 survey respondents who did not teach sustainability in their mechanical engineering undergraduate courses stated that a sustainability curriculum for their specific subject area would encourage them to incorporate sustainability into one or more courses. And 13 of the 53 survey respondents who did teach sustainability in their undergraduate mechanical engineering courses said that a prepared curriculum was helpful to them when they were first integrating sustainability. Two of the interviewees reiterated this sentiment that a prepared curriculum was helpful to them as they were first incorporating sustainability into their curricula. Interviewee 3 articulated a need for a prepared curriculum: "There weren't many textbooks available. So, it took pulling stuff from all over the place. I definitely did not have a whole lot of material to pull from. I had to develop a lot of stuff on my own". Curriculum examples include the Engineering for One Planet initiative [39] and the Center for Sustainable Engineering Curriculum Modules [40].

A desire for **training** was evident in both the survey and interview results. Five of the 14 survey respondents who did not include sustainability in their undergraduate engineering courses said that a workshop or training would encourage them to incorporate sustainability into one or more of their courses. And six of the 53 survey respondents with sustainability in their curricula said that a workshop and/or training was helpful to them when they were first integrating sustainability. Interviewee 9 said "I have the training" when describing her ability to include sustainability in her courses, demonstrating confidence and empowerment. Interviewees 3 and 5 expressed an initial lack of training or expertise in sustainability. Interviewee 3 said, "I'll even tell the students: 'I am not an expert in this conversation, but let's have the conversation.'"

3.4.3. Organization

The organizational aspects that emerged as needs were space in a curriculum and consistency in defining sustainability. A **crowded curriculum** was identified as a hindrance. Eight of the 14 survey respondents who did not include sustainability in their undergraduate engineering courses said that there were too many other topics to cover in the course and that this was a reason why they had not included sustainability in their courses, and one stated that "no time" was a reason. Interviewee 2 asked, "How do you incorporate all these things into your classes with the core curriculum that you need to teach, right? And how do you have time?" And thus, the presence of space within a four-year plan and/or within a specific course's content is a need that professors face when exploring the option of adding sustainability into an existing course or adding a dedicated sustainability course.

Knowing what to teach concerning sustainability can constitute a challenge, as there is a lack of **consistency** in opinions on the key knowledge and competencies engineering students should learn about sustainability. A focus group participant observed, "When I started out trying to figure out for myself what sustainability meant, I realized that people use that word in a million different ways…and there's really no broad consensus about it". And Interviewee 5 stated, "It is really hard to know what should be in a sustainability course…. I think people may think of sustainability as not so much like a discipline or a topic, as it is something kind of equivalent to critical thinking". Interviewee 5 went on to describe the standards for what would be included in a thermodynamics course and explained that there would be consistency across programs and universities and that this is not the case with sustainability. Professors may find it helpful if there were a standard or consistent curriculum for sustainability courses or sustainability inclusion principles.

3.4.4. Meeting Student Needs

Meeting the needs of their students was cited as a motivator by some of the faculty members. These can be short-term student needs (e.g., providing motivation and support for learning and developing them as whole people) and long-term needs as engineering professionals who, we hope, design for sustainability. Most of the interviews focused on the short-term needs of students as learners and as whole people, rather than specific needs in their future careers as engineers. Based on their actions, it could be presumed that all the interviewees believed sustainability awareness would be beneficial to students in their futures.

Student interest in the topic of sustainability was identified as a motivator by four interviewees and five survey respondents. Interviewee 3 said, "I think they (i.e., the students) definitely find the sustainability stuff interesting". Interviewee 4 described the intrigue experienced by most students in a manufacturing case study concerning the popular toy Lego. In the survey, two respondents indicated student interest or motivation as something that inspired them to incorporate sustainability into their undergraduate courses (e.g., "I believe it is a motivator for students"), and three additional responses to the final open-ended question noted student interest (e.g., "Student interest has led the way"). Students may be among the biggest proponents of sustainability's integration; in a study with over 1700 university students, almost all the students agreed that higher education institutions should actively incorporate and promote sustainable development [41].

Relating to real-world content can pique student interest. All ten interviewees described bringing sustainability issues into their classes through real-world examples, case studies, and scenarios. This included reading assignments concerning current events as part of homework assignments alongside solving technical problems, playing games that embedded sustainability principles, and in-class discussions. For example, Interviewee 6 said, "I take a minute to even recognize what's going on in the world…that we're a part of the world, not separate from it…I don't separate the learning about sustainability from the participation in a planet of 8 billion people". Interviewee 4 asked her class "why" questions concerning current events that had links to engineering. Interviewee 2 said, "I do an activity where I show them quotes and then they get to disagree or agree with it. And that leads to a nice conversation about different aspects of sustainability. …. I love how they transform by understanding the culture, while suddenly studying the mathematics at the same time". Interviewee 8 described playing a game centered around manufacturing a product with limited time and money, factoring in mining for the rare earth metals needed. The activity included reflecting on their personal environmental impact, as well as designing for recyclability or repairability. Interviewee 1 said, "over the years I've actually added a lot more realism to my class… We always have a climate week…. we integrate politics into things, and discuss public perception and marketing … I added all that in there because I felt that my students were so focused on the math and the design that they weren't understanding that, hey, you gotta look at the big picture". Previous research has found that real-world contexts positively impact learning due to learner motivation [42,43].

Some classes are more challenging to integrate sustainability into. Interviewee 4 discussed the difficulty of integrating sustainability topics into statics, contrasted with the relative ease in her manufacturing course. However, Interviewee 3 included sustainability in statics by connecting the base of a problem to something real in sustainability and then using that as an opportunity for discussion: for example, completing statics calculations on a nuclear energy gantry crane and then talking about energy options.

Acknowledging that each student is a **whole person** with emotions came up as important in meeting student needs concerning sustainability education. Interviewee 6 said, "To really process any and all of the meaning of all the sustainability stuff, you need your whole body…You need to be able to go, that feels scary. How do I deal with my fear? …. We're talking about people's futures and the life they imagine for themselves…. You can precipitate an existential crisis…" Being cognizant of that is important to student well-being. Interviewee 10 focused on framing sustainability from a place of hope and empowerment: "I've tried to always bring it to students in a way, you know, we can do something about this, you can do something about this". Interviewee 3 also shared this perspective of trying to keep students motivated despite the gloom.

Thinking about students' emotions is quite different from the perspective of one of the survey respondents who did not teach sustainability: "It would be much easier to incorporate sustainability topics if there were well-established, objective curriculum materials that treated sustainability concepts with the rigor we want for an engineering course (by that, I'm thinking of ways to assess student learning of sustainability concepts via exams that include quantitative problems)". Thus, the best way to teach sustainability as viewed by innovators and early adopters perhaps differs from the perspective of later adopters.

3.5. Adopter Group Synthesis

The influences on the faculty integration of sustainability into ME courses that presented themselves in this research were cross-referenced with the suggestions for different adopter categories described in diffusion of innovation theory [18,19], as shown in Figure 1. Aligned with DoI recommendations, the results of this study indicated that **early adopters** could be motivated by providing incentives like performance evaluations and/or financial bonuses for new and innovative ideas in sustainability integrations or for spreading their innovative sustainability practices to other professors, courses, or programs. Second, a mentorship program and other peer-to-peer communities could provide strong support and feedback. Early adopters may also appreciate public praise as front runners and innovators. Grant opportunities for funding to keep pushing the frontiers and taking sustainability to the next level could be of interest.

For faculty members with dispositions in the **early majority**, both training and a prepared curriculum could maximize the ease and simplicity of incorporating sustainability into their courses. A university sustainability office could provide "customer service and support". Support could also be provided via peer-to-peer channels. If a university and/or college requires a certain number of courses per major or adopts strategic goals related to sustainability (such as committing to earn the AASHE STARS platinum rating), this promotes the social norm of sustainability education, which may spur **late-majority faculty**. Student interest and peer-to-peer communities could also promote a new social norm of sustainability's importance in ME courses and curricula. Aligned with this idea of social norms, in the focus group, it was asked, "do you think mechanical engineers think that sustainability is a part of their job?" Unanimous head shaking occurred, signifying "no". Another encouragement for late-majority faculty would be to make sustainability integration highly convenient by having a standard set of curricula or definitions on what sustainability inclusions look like and/or pre-existing integration into the textbooks that they are already using.

It is worth noting that particular ME departments may have a range of adoption levels among their faculties, such that different user groups are locally relevant. For example, a faculty member in an ME department without sustainability integration in any courses would be a local innovator, but by looking beyond their own department, they might be part of a late majority at their institution. For example, a focus group participant stated, "I would say that the majority of the sustainability thinking, I think happens outside of engineering". In addition, particular types of sustainability topics or teaching approaches may be locally innovative and still involve global peers who are willing to mentor, share information, or publish about their teaching practices.

Faculty User Group	DOI suggestions to meet user needs	This study's suggestions to meet user needs	
Innovators	Provide support and publicity for their ideas, invite them to be design partners	Peer to Peer	
Early Adopters	Promote them as leaders, offer strong face-to-face support, make idea more convenient and low cost, provide regular feedback	Incentive Mentorship program	Peer to Peer
Early Majority	Redesign to maximize ease and simplicity, provide strong customer service and support, lower cost of entry	Training Prepared curriculum Sustainability office	
Late Majority	Promote social norms, keep refining for increased convenience and reduced costs, emphasize risks of being left behind	Bandwidth Consistency Meet student needs Supporting policy	
Laggards	Maximize familiarity with the innovation, offer high level of personal control over when, where, how, and whether they accept the new innovation		

Figure 1. User-based needs from DoI [2,3] aligned with the results obtained in this study.

4. Summary and Conclusions

In this study, the factors found to bolster front runners' sustainability teaching practices in ME courses included publications, training, a prepared curriculum, and a variety of other supports. Of those who were interviewed, all ten were motivated by perceived student interest in sustainability. Three interviewees described how they empowered their students by helping them see that they could make a difference and be one piece of the complex puzzle of working toward a more sustainable future. Half of those who were interviewed shared personal experiences that were significant motivators in their commitment to sustainability education. Additionally, peer-to-peer interactions and institution's missions or policies were both factors that influenced four of the interviewed front runners. For example, two of the interviewees worked for higher education institutions that mandated sustainability courses for all students, thereby setting a social norm of sustainability inclusion. Significantly, these two professors had the highest confidence in student awareness of sustainability upon graduation versus those teaching at institutions without mandatory sustainability inclusion.

Among the 53 survey respondents who taught sustainability, 85% indicated that a personal passion was among their reasons for teaching sustainability, followed by 32% citing an institution mission or policy, 26% acknowledging the importance of a prepared curriculum (thus increasing convenience), and 25% the importance of having a mentor (providing face-to-face support and regular feedback). Reading publications, training, and student interest were additional motivators and supports.

The factors that could spur non-adopters to integrate sustainability into their courses, as reported by 14 survey respondents, included a prepared curriculum ($n = 6$), incentives ($n = 6$), and training ($n = 5$). The inhibitors of sustainability's integration included too many other topics to cover in their courses ($n = 8$).

This research identified numerous suggestions to help propagate sustainability education among more instructors in mechanical engineering. The results listed here should not be considered exhaustive due to the limited number and characteristics of the interview and survey participants, but they still provide useful suggestions. All instructors can have a role to play in educating engineers to contribute to a sustainable future. And diffusion of innovation theory may provide helpful lessons to reach instructors who are not teaching sustainability in mechanical engineering and other disciplines.

Author Contributions: Conceptualization, J.K.T. and A.R.B.; methodology, J.K.T. and A.R.B.; validation, J.K.T. and A.R.B.; analysis, J.K.T. and A.R.B.; investigation, J.K.T. and A.R.B.; resources, A.R.B.; data curation, J.K.T. and A.R.B.; writing—original draft preparation, J.K.T.; writing—review and editing, A.R.B.; supervision, A.R.B.; project administration, J.K.T. and A.R.B. All authors have read and agreed to the published version of the manuscript.

Funding: This research received no external funding.

Institutional Review Board Statement: This study was conducted in accordance with the University of Colorado Boulder Institutional Review Board, protocol number 22-0545, approved in October 2022.

Informed Consent Statement: Informed consent was obtained from all subjects involved in the study.

Data Availability Statement: The survey data presented in this study are available on request from the corresponding author. The interview participants did not consent to their data being shared.

Acknowledgments: The authors would like to thank all the survey respondents, interviewees, and focus group participants.

Conflicts of Interest: The authors declare no conflict of interest.

References

1. Corres, A.; Rieckmann, M.; Espasa, A.; Ruiz-Mallén, I. Educator Competences in Sustainability Education: A Systematic Review of Frameworks. *Sustainability* **2020**, *12*, 9858. [CrossRef]
2. Serafini, P.G.; Morais de Moura, J.; Rodrigues de Almeida, M.; Dantas de Rezende, J.F. Sustainable Development Goals in Higher Education Institutions: A systematic literature review. *J. Clean. Prod.* **2022**, *370*, 133473. [CrossRef]
3. Thurer, M.; Tomasevic, I.; Stevenson, M.; Qu, T.; Huisingh, D. A systematic review of literature on integrating sustainability into engineering curricula. *J. Clean. Prod.* **2018**, *181*, 608–617. [CrossRef]
4. Tejedor, G.; Segalàs, J.; Rosas-Casals, M. Transdisciplinarity in higher education for sustainability: How discourses are approached in engineering education. *J. Clean. Prod.* **2017**, *175*, 29–37. [CrossRef]
5. Gutierrez-Bucheli, L.; Kidman, G.; Reid, A. Sustainability in engineering education: A review of learning outcomes. *J. Clean. Prod.* **2022**, *330*, 129734. [CrossRef]
6. Filho, W.L.; Frankenberger, F.; Salvia, A.L.; Azeiteiro, U.; Alves, F.; Castro, P.; Will, M.; Platje, J.; Lovren, V.O.; Brandli, L.; et al. A framework for the implementation of the Sustainable Development Goals in university programmes. *J. Clean. Prod.* **2021**, *299*, 126915. [CrossRef]
7. Tisdale, J.; Bielefeldt, A. Sustainability in Mechanical Engineering Undergraduate Courses at 100 Universities. *ASME Open J. Eng.* **2023**, *2*, 021049. [CrossRef]
8. Lattuca, L.; Stark, J. *Shaping the College Curriculum*; Jossey-Bass: San Francisco, CA, USA, 2009.
9. Menon, M. Understanding Faculty Decision-Making in Engineering Education of Sustainable Development. Ph.D. Thesis, Virginia Polytechnic Institute and State University, Blacksburg, VA, USA, 2023.
10. Garibay, J.; Vincent, S.; Ong, P. Diversity content in STEM? How faculty values translate into curricular inclusion unevenly for different subjects in environmental and sustainability programs. *J. Women Minor. Sci. Eng.* **2020**, *26*, 61–90. [CrossRef]
11. Akins, E.I.; Giddens, E.; Glassmeyer, D.; Gruss, A.; Kalamas Hedden, M.; Slinger-Friedman, V.; Weand, M. Sustainability Education and Organizational Change: A Critical Case Study of Barriers and Change Drivers at a Higher Education Institution. *Sustainability* **2019**, *11*, 501. [CrossRef]

12. Reynante, B. Engineering for One Planet Literature Review Report 2022. Available online: https://engineeringforoneplanet.org/wp-content/uploads/EOP-Systems-Change-Literature-Review-Report-Final.pdf (accessed on 30 March 2023).
13. Currie, G.; Henderson, A.; Hoult, R. Diffusion of innovation in an Australian engineering school. *Australas. J. Eng. Educ.* **2021**, *26*, 219–226. [CrossRef]
14. Menzli, L.J.; Smirani, L.K.; Boulahia, J.A.; Hadjouni, M. Investigation of open educational resources adoption in higher education using Rogers' diffusion of innovation theory. *Heliyon* **2022**, *8*, e09885. [CrossRef] [PubMed]
15. Scott, S.; McGuire, J. Using Diffusion of Innovation Theory to Promote Universally Designed College Instruction. *Int. J. Teach. Learn. High. Educ.* **2017**, *29*, 119–128.
16. Bennett, J.; Bennett, L. A review of factors that influence the diffusion of innovation when structuring a faculty training program. *Internet High. Educ.* **2003**, *6*, 53–63. [CrossRef]
17. Diffusion of Innovations Summary and Review. LifeClub. 2019. Available online: lifeclub.org/books/diffusion-of-innovations-everett-m-rogers-review-summary (accessed on 30 March 2023).
18. Rogers, E. *Diffusion of Innovations*; Free Press: New York, NY, USA, 2003.
19. Robinson, L. A Summary of Diffusion of Innovations. 2009. Available online: https://twut.nd.edu/PDF/Summary_Diffusion_Theory.pdf (accessed on 30 March 2023).
20. Diffusion of Innovation. Behavioral Change Theories. 2022. Available online: https://sphweb.bumc.bu.edu/otlt/mph-modules/sb/behavioralchangetheories/behavioralchangetheories4.html (accessed on 30 March 2023).
21. Sahin, I. Detailed Review of Rogers' Diffusion of Innovations Theory. *Turk. Online J. Educ. Technol.* **2006**, *5*, 14–23.
22. Bielefeldt, A.; Polmear, M.; Knight, D.; Canney, N.; Swan, C. Disciplinary variations in ethics and societal impact topics taught in courses for engineering students. *J. Prof. Issues Eng. Educ. Pract.* **2019**, *145*, 04019007. [CrossRef]
23. ASEE. *Profiles of Engineering and Engineering Technology*; American Society for Engineering Education: Washington, DC, USA, 2023.
24. Enelund, M.; Wedel, M.; Lundqvist, U.; Malmqvist, J. Integration of education for sustainable development in a mechanical engineering programme. In Proceedings of the 8th International CDIO Conference Queensland University of Technology, Brisbane, Australia, 1–4 July 2012.
25. Ramanujuan, D.; Zhou, N.; Ramani, K. Integrating environmental sustainability in undergraduate mechanical engineering courses using guided discovery instruction. *J. Clean. Prod.* **2019**, *207*, 190–203. [CrossRef]
26. Issa, R. Teaching sustaiinability in mechanical engineering curriculum. *Athens J. Technol. Eng.* **2017**, *4*, 171–189.
27. Jha, N. Engineering Mechanics and Sustainable Engineering. In Proceedings of the 10th Engineering Education for Sustainable Development Conference, Cork, Ireland, 14–16 June 2021.
28. Hollander, R.; Amekudzi-Kennedy, A.; Bell, S.; Benya, F.; Davidson, C.; Farkos, C.; Fasenfest, D.; Guyer, R.; Hjarding, A.; Lizotte, M.; et al. Network priorities for social sustainability and education: Memorandum of the Integrated Network on Social Sustainability Research Group. *Sustain. Sci. Pract. Policy* **2016**, *12*, 16–21. [CrossRef]
29. Menon, M.; Katz, A.; Paretti, M.C. A thematic and trend analysis of engineering education for sustainable development. In Proceedings of the 2022 ASEE Annual Conference & Exposition, Minneapolis, MN, USA, 26–29 June 2022.
30. Davidson, C.; Heller, M. Introducing sustainability into the engineering curriculum. In *ICSI 2014: Creating Infrastructure for a Sustainable World*; American Society of Civil Engineers: Reston, VA, USA, 2014. [CrossRef]
31. Tisdale, J.K. Inclusion of Sustainability in Mechanical Engineering Courses: A Synthesis of Current Practices and Lessons Learned. Ph.D. Thesis, University of Colorado, Boulder, CO, USA, 2023.
32. National Science Foundation. Awards Simple Search. Available online: https://www.nsf.gov/awardsearch/ (accessed on 30 March 2023).
33. Saldana, J. *The Coding Manual for Qualitative Researchers*; Sage Publications Ltd.: London, UK, 2009.
34. Long, Z.; Buzzanell, P.; Kokini, K.; Wilson, R.; Batra, J.; Anderson, L. *Exploring Women Engineering Faculty's Mentoring Networks*; American Society for Engineering Education (ASEE) Annual Conference & Exposition: Atlanta, Georgia, 2013.
35. AASHE. Campus Sustainability Hub: Academic Programs. Available online: https://hub.aashe.org/browse/types/academicprogram/ (accessed on 30 March 2023).
36. Willats, J.; Erlandsson, L.; Molthan-Hill, P.; Dharmasasmita, A.; Simmons, E. *A University Wide Approach to Embedding the Sustainable Development Goals in the Curriculum—A Case Study from the Nottingham Trent University's Green Academy*; Sustainability Series; Springer: Berlin/Heidelberg, Germany, 2017.
37. University of Michigan-Dearborn. Faculty Mentoring. University of Michigan-Dearborn. 2023. Available online: https://umdearborn.edu/faculty-senate/faculty-mentoring. (accessed on 6 May 2023).
38. Dumitru, D.E. Reorienting higher education pedagogical and professional development curricula toward sustainability—A Romanian perspective. *Int. J. Sustain. High. Educ.* **2017**, *18*, 894–907. [CrossRef]
39. Engineering for One Planet (EOP). 2023. Available online: https://engineeringforoneplanet.org/ (accessed on 30 March 2023).
40. Georgia Institute of Technology. Sustainable Engineering Education. 2023. Available online: https://research.gatech.edu/sustainability/education (accessed on 30 March 2023).
41. Alexio, A.M.; Leal, S.; Azeiteiro, U.M. Higher education students' perceptions of sustainable development in Portugal. *J. Clean. Prod.* **2021**, *327*, 129429. [CrossRef]

42.	Duffy, T.; Cunningham, D. Constructivism: Implications for the Design and Delivery of Instruction. In *Handbook of Research on Educational Communications and Technology*; Simon & Schuster: New York, NY, USA, 1996.
43.	Lunce, L. Simulations: Bringing the benefits of situated learning to the traditional classroom. *J. Appl. Educ. Technol.* **2006**, *3*, 37–45.

 *sustainability*

Article

A Study of Safety Issues and Accidents in Secondary Education Construction Courses within the United States

Tyler S. Love [1,*] and Kenneth R. Roy [2]

1   Department of the Built Environment, University of Maryland Eastern Shore, Baltimore, MD 21230, USA
2   Department of Environmental Health & Safety, Glastonbury Public Schools, Glastonbury, CT 06033, USA; safesci@sbcglobal.net
*   Correspondence: tslove@umes.edu

Abstract: Hands-on learning is paramount to teaching concepts about construction and the built environment; however, this poses some inherent safety risks. This study analyzed a subsample of 119 teachers from a national safety study, focusing on those who taught secondary-level construction courses. The current study aimed to examine the demographics of construction teachers, accident occurrences in construction courses compared to other secondary-level technology and engineering education (TEE) courses, and safety factors and items associated with accident occurrences in construction courses. The analyses revealed that a significantly higher number of minor accidents occurred in construction courses compared to other TEE courses during a five-year span. Additionally, 20 safety factors were found to be significantly associated with increases or decreases in accident occurrences. Most notably, increases in major accident occurrences increased with marginal significance when average class sizes (occupancy load) exceeded 20 students. Construction courses were also found to have significantly more accidents involving hand and power tools compared to other TEE courses. This research contributes to the limited literature on this topic and has implications for proactively limiting potential safety hazards and resulting risks. It also provides data to inform the safety efforts of post-secondary construction programs and the construction industry.

Keywords: construction education; built environment; occupational safety and health; risk management; overcrowding; safety training; career and technical education; technology and engineering education; teacher preparation; alternative certification

 check for updates

Citation: Love, T.S.; Roy, K.R. A Study of Safety Issues and Accidents in Secondary Education Construction Courses within the United States. *Sustainability* **2023**, *15*, 11028. https://doi.org/10.3390/su151411028

Academic Editors: Clara Viegas and Natércia Lima

Received: 15 June 2023
Revised: 2 July 2023
Accepted: 12 July 2023
Published: 14 July 2023

1. Introduction

There is an alarming shortage of highly skilled and trained workers in the construction industry within the United States (U.S.) [1]. It has been projected that employment opportunities in the construction industry will increase a total of 4.4% from 2020 to 2030. Federal support from the Infrastructure Investment and Jobs Act is expected to help create 1.5 million construction jobs [2]. However, the average age of construction workers continues to increase, meaning there will be an increasing demand for younger skilled workers to replace aging employees as they retire or pursue other opportunities. Construction laborers, carpenters, and electricians are projected to be the occupations with the highest demand in the construction industry until 2030. Additionally, employment of solar installers, construction managers, and telecom-line installers is projected to grow the fastest until 2030 [2]. Each of these occupations has unique and inherent safety risks, which is why high-quality training is crucial for the construction industry. Reports show that construction is one of the most hazardous industries in the U.S., with parts/materials, tools/instruments/equipment (especially ladders), and falls/slips/trips being among the most frequent cause or items associated with nonfatal construction injuries. Moreover, minors and young workers have higher nonfatal injury rates than other age groups in the construction industry [3,4]. Teaching minors and young workers to have a better understanding of construction safety practices can help reduce these injuries and drastically improve the lives of those working in the construction industry [3].

Developing safer construction habits begins with proper safety training and modeling appropriate safety practices during secondary technology and engineering education (TEE) and career and technical education (CTE) courses. As secondary TEE and CTE educators aim to enhance students' technological and engineering literacy, and help students earn industry recognized credentials to be prepared for entering the workplace and post-secondary programs, safety must remain a core focus [5]. Safety is not only a critical concern to construction industry employers, it is also a core component of K-12 TEE academic standards which guide curriculum, instruction, and assessment. *The Standards for Technological and Engineering Literacy* [6] places a strong emphasis on safety throughout the standards, practices, and context areas [5,7]. Additionally, the *Association for Career and Technical Education (ACTE) Quality CTE Program of Study Framework* [8] highlights safety as one of the 12 core elements needed for a high-quality CTE program. Element 7 specifically focuses on safer facilities, equipment, processes, and practices that align with industry expectations related to the program of study. Emphasizing and modeling safer practices during secondary TEE and CTE courses has important implications for improving young worker safety (including young workers in the construction industry), safety in the workplace, and safety knowledge and awareness of students preparing to enter post-secondary construction programs [5,9]. Enhancing the safety of students and young workers can also help improve their professional safety practices and resulting safety outcomes, contributing to greater efficiency and sustainability among the construction industry [10]. Therefore, the main goal of this study was to improve safety in secondary TEE and CTE construction courses by examining the safety issues and accident occurrences reported by instructors of these courses.

2. Literature Review

2.1. Teaching Construction through a Technological and Engineering Literacy Lens

Construction concepts are taught in a broad range of secondary education curricula, courses, and programs using a variety of pedagogical strategies. One perspective is that foundational construction technology (also referred to as the built environment) concepts and an overview of the construction process should be taught to all students to help them become more technologically and engineering-literate citizens [6,11]. Blankenbaker's [12] textbook, written for secondary TEE programs interested in teaching a course dedicated to construction processes and procedures, focuses on the following topics: Construction tools/equipment and safety, construction materials, project design, construction management, site preparation, superstructures, mechanical/electrical/plumbing systems, finishing steps of a project, remodeling, commercial and industrial construction applications, and construction careers. However, as Love and Salgado [11] highlighted, instruction about construction concepts is often relegated to a broad single unit within secondary TEE courses that are available to the general secondary student population. In alignment with the technological and engineering literacy focus of K-12 TEE standards [6], Love and Salgado proposed that, "the broad overview of construction and structural technologies should be taught from a design-based approach, advocating for students to have an understanding of the construction process and corresponding science, technology, engineering, and mathematics (STEM) concepts" [11] (p. 22). Construction concepts would fall within the built environment context area of the STEL and should be taught in a hands-on, design-based manner that allows students "opportunities to safely design, use, and assess structures and materials" [6] (p. 109). Furthermore, instruction related to the built environment should help students to apply core technology and engineering (T&E) concepts, T&E practices, and make cross-cutting connections to other T&E context areas and STEM disciplines. For example, students can evaluate and test the properties of materials used in a scaled structure, or they could study heat loss in a structure and devise ways to optimize energy use [6] (p. 110).

Educators and researchers have provided cross-cutting examples in which students utilize microcomputers and sensors to design and program prototypes of scaled smart home devices (motion activated garage door, security systems, CO_2 sensors, and temperature feedback systems, etc.). These examples illustrate how students could apply engineering design skills, sustainability concepts (impacts of technology), systems thinking practices, and optimism practices [6], within the built environment and computation, automation, artificial intelligence, and robotics context areas of the STEL. Love and Salgado [11] also provided numerous examples of ways in which educators could embed design-based construction lessons within technological and engineering literacy-focused curricula.

2.2. Teaching Construction within Secondary Pre-Engineering Education Courses

Another perspective is that more in-depth construction and built environment concepts should be taught within the context of pre-engineering education courses offered at the secondary level. These courses are generally geared toward students who hope to pursue a degree from a post-secondary engineering-related program. One popular example in secondary U.S. schools is Project Lead the Way's (PLTW) Civil Engineering and Architecture course [13]. This focus is also evident from some secondary education course textbooks. Brown et al. [4] demonstrate connections to the construction industry and the built environment across multiple engineering fields, such as materials engineering, civil engineering, environmental engineering and others. While there are subtle differences, pre-engineering and broader technological and engineering literacy courses both use 3D modeling software to help students safely design, test, and optimize their theoretical solutions before testing them through experiences with practical materials in a lab setting. Hughes and Merrill presented a series of examples for secondary TEE and pre-engineering educators to teach statics concepts through the design and testing of a concrete beam [14,15], the creation and testing of trusses from craft sticks [16], the testing of an aluminum can crusher [17], and other practical design challenges. While all these examples provide practical hands-on applications for students to see authentic construction and built environment theories in action, they also must be conducted with safety in mind.

2.3. Construction as a CTE Program of Study

Instruction pertaining to construction and the built environment also falls under the architecture and construction career cluster within CTE programs in local school systems. This career cluster is "focused on careers in designing, planning, managing, building and maintaining the built environment" [18] (para. 1). It encompasses programs focused on various facets of the construction industry and the built environment (carpentry, masonry, project management, alternative energy systems, etc.). Many secondary programs in the architecture and construction career cluster work to align with the accreditation standards set forth by the National Center for Construction Education and Research (NCCER). In addition to overseeing industry-recognized standards for safer and more rigorous construction instruction, NCCER provides curricular and assessment resources for school systems, training for instructors, career and articulated post-secondary education credit pathways, career starter resources, and other resources to help instructors provide high-quality instruction that meets the expectations of construction industry employers. An example of this can be seen at Roxbury High School in Succasunna, New Jersey, USA. In their Structural Design and Fabrication course students design and build a modular house that is donated to a family through their local Habitat for Humanity program. From this experience, students learn various aspects of designing and constructing a residential structure from start to finish. Students also complete the Occupational Safety and Health Administration (OSHA) 10-h outreach safety training program focused on construction standards. Safety is a core focus in everything that they do, as demonstrated by their 2022 National Association of Home Builders (NAHB) Safety Award for Excellence [19].

As Love and Roy [5] discussed, the lines can often become blurred between technological and engineering literacy programs and CTE programs of study within various states in the U.S. They highlighted that in many states, TEE programs fall under the supervision of the state department of education's CTE division. One unique example is in the state of Maryland, where TEE courses are classified like other instructional content areas; however, PLTW courses (including civil engineering and architecture) are classified under the manufacturing, engineering, and technology CTE career cluster [20]. In other states like Virginia and Wisconsin, TEE programs are classified as a distinct CTE program area by their state department of education. In addition to the expansive content knowledge, pedagogical knowledge, and safety practices needed to teach specialized courses related to various TEE and CTE programs, the critical shortage of highly-qualified teachers in these areas has increased safety concerns and risks [7,21–23].

2.4. Recruiting, Preparing, and Retaining Secondary Education Construction Teachers

Trends have indicated that across the U.S. there is a growing shortage of TEE [24] and CTE teachers [25]. These shortages have been linked to various reasons, including burnout following the COVID-19 pandemic, large numbers of teachers from the baby boomer generation retiring with fewer young people to fill these vacancies, and lower salaries compared to working in business or industry. For example, in 2021 the average salary of TEE and CTE teachers with a bachelor's degree was $67,610, whereas the average salary of a construction manager with a bachelor's degree was $112,790 [26]. The growing teacher shortage has increased the number of TEE and CTE teachers being hired with alternative or provisionary state teaching certification [7,21,23,25,27,28]. For example, educators teaching TEE courses in the U.S. with alternative certification rose from 24% in 2008 to 39% in 2017 [27]. As researchers have cautioned, alternative certification can pose some serious safety concerns [7,21,23]. Even for individuals entering the teaching profession with years of experience from industry, overseeing the safety of young novice students in a secondary school environment compared to skilled adults in an industry setting requires a unique skillset. Highly qualified educators that complete a teacher preparation program often receive specialized coursework in classroom management, facilities planning, and laboratory safety [29]. Additionally, most TEE teacher preparation programs require students to complete a course on construction technologies, a course on manufacturing technologies, and a teaching methods course [29] which should all incorporate content related to safety within these technical contexts. Duncan et al. [30] found that, in comparison to educators who entered CTE through an occupational teacher preparation route, those who completed a traditional CTE teacher preparation program reported being more efficacious in teaching proper safety practices in the lab and teaching proper safety attitudes in CTE courses.

Researchers have also expressed concerns about brief professional development (PD) preparation models, such as that used by PLTW to quickly prepare educators (including those with no degree or certification in a TEE- or CTE-related area) to teach TEE and CTE courses involving potentially hazardous laboratory activities [22,24]. In some states, like Maryland, professionals with technical expertise and work experience from industry can earn their initial certification through a four-course pathway. State regulations require TEE and CTE educators entering the profession through this route to complete coursework focused on laboratory safety practices, classroom management, instructional strategies, assessment, and safer teaching and learning strategies for students of all abilities. While this is still a limited experience in comparison to a traditional four-year teacher preparation program, it helps educators combine their expertise from industry with the special pedagogical practices needed to facilitate and supervise safer TEE and CTE teaching and learning experiences.

The shortage of highly qualified TEE and CTE teachers has also led school systems to ask certified TEE and CTE teachers to take on additional responsibilities, such as teach a wider variety of courses. Fifty-one percent of TEE and CTE teachers in Love and Roy's [5] national safety study reported that, on average, they have more than three different courses they are tasked with preparing for and teaching each semester. This elicits concerns as more than three course preps in a semester can put additional responsibilities on instructors (especially new and alternatively certified educators) and has been linked to significant increases in accident occurrences [31,32]. Previous studies have demonstrated that the aforementioned safety risk factors have been found to significantly increase the odds of an accident in TEE and CTE courses; however, certain safety protective factors and practices described in the next section have been found to help reduce the effect of prevalent safety risk factors [31,32].

2.5. Safety: A Critical Concern in Secondary Construction Courses and Industry

TEE and CTE teachers must be prepared to face an amalgam of potential safety issues due to the inherently hazardous nature of these courses and programs [5]. When focusing on construction and the built environment courses, there are a number of unique safety hazards that differ from other courses which educators need to be trained to properly address. These safety issues include designing with safety in mind, safety ethics considerations, environmental considerations, safety considerations for those of varying abilities, codes (e.g., building codes), product standards, state and federal occupational safety and health (OSH) standards, engineering controls, personal protective equipment (PPE), body mechanics (e.g., lifting), ladder and scaffolding safety, electrical safety, hand/power tools/equipment safety, fire hazards and extinguishers, hazardous materials, heat exposure, and other pertinent safety issues for the task at hand [12]. In addition to live instructor demonstrations, safety testing, and direct supervision from the trained educator on site, high school students can enhance their safety knowledge and practices through earning their industry recognized OSHA 10 cards in general industry safety standards (29 CFR Part 1910) and construction safety standards (Part 29 CFR Part 1926). Both safety standards would apply to secondary construction courses and the construction workplace as described in the discussion section of this article. Recognizing the importance of safety in preparing students for the construction industry and post-secondary programs, NCCER has partnered with CareerSafe™ to embed safety modules into their secondary education curriculum [33]. This provides students with an industry-recognized credential and additional safety knowledge that employers value. Furthermore, construction industry employers have expressed the value of prospective employees (e.g., students) receiving First-Aid, CPR, and AED certification because of how important it is in the event of an accident (e.g., electrical shock, impalement, lacerations, etc.).

2.6. Previous Research on Safety in Secondary Construction Courses

For decades the literature has documented safety as a top concern among TEE and CTE teachers and administrators [9,21,34–37]. This is not surprising due to the inherently hazardous nature of the hands-on learning experiences provided in these programs to better prepare students for the experiences they will encounter in the workplace and in post-secondary programs. While there is an abundant amount of literature examining various aspects of safety in the construction industry, the amount of research focused on safety in secondary education construction programs is scarce, especially within the U.S. [38].

There have been a few international studies that provide valuable insight for research investigating safety in secondary education construction programs. Grytnes et al.'s [39] study examined Swedish and Danish students completing a workplace apprenticeship as part of their secondary education construction program. They found that students who participated in company-based safety training from a student role had greater perceptions of the safety climate at that company in comparison to students who were treated more like an employee. It was discovered that students who were treated like employees were more hesitant to voice concerns or ask questions about safety in fear of losing their apprenticeship, whereas those treated like students viewed the apprenticeship more like a learning experience. Grytnes et al. concluded that there needs to be a careful balance of safety training and oversight between secondary construction programs and apprenticeship hosts, with secondary construction teachers serving as an advocate to help students feel more comfortable about enhancing their safety knowledge and skills in construction workplace settings. Other insightful international studies have focused on the importance of OSH training for secondary-level CTE programs. One Canadian study [40] found that CTE teachers preferred to teach OSH content with which they were more familiar, and required training on how to incorporate OSH content into their courses so that content was not viewed as simply another element to be covered within the curriculum. Chatigny et al. [40] concluded that CTE teachers need the appropriate training, knowledge, teaching and learning modality resources, and opportunities to provide concrete learning experiences to allow for the implementation and development of OSH skills. Moreover, Nykänen et al.'s [41] study of Finnish secondary CTE students found that OSH attitude training significantly increased the students' safety preparedness, increased their internal safety locus of control, and reduced their risk-taking attitudes.

Within the U.S., Jones et al. [42] conducted a Delphi study involving a 12-person panel comprised of secondary and post-secondary construction teachers, recent graduates from construction or architecture programs, and professionals who worked in the industry. The panel members identified OSHA safety as a dominant technical competency required to effectively teach construction and architecture at the secondary and post-secondary levels. Similarly, CTE educators in Lupton's [36] study identified providing for student safety as the most important skill needed to teach in a CTE program. Nichols [43] found that approximately 23% of residential construction teachers in high school vocational education settings were involved in an accident over a five-year span. Additionally, 42% of the teachers indicated a student was involved in an accident in their courses, and 8% reported having a student sustain a serious injury during that five-year span. Bush et al. [38] discovered that secondary construction teachers felt overwhelmed by the responsibilities of meeting both educational and industry standards, and the limited time they had to adequately prepare for integrating high-quality OSH training in their courses. Specifically, they found that the amount of required information that needed to be taught for students to earn their OSHA 10 card was very challenging to integrate in a packed construction curriculum. Another challenge raised by the construction educators in Bush et al.'s study was the broad range of students taught by instructors. They expressed how challenging it can be to provide safer modifications and accommodations for each student in a full class that is engaged in hands-on construction activities. Construction teachers emphasized the need for classroom-ready OSH teaching resources that also provided some flexibility for the instructor to adapt or modify for their students' needs [38].

Threeton and Evanoski [9] examined the safety of CTE teachers and school districts from thirty counties in Pennsylvania and found many alarming safety concerns. Specific to carpentry and masonry teachers within their study, they found: 39% did not have a mandated written safety policy developed by their school district, 87% did have a written safety policy developed by their program, 25% did not have the health and safety elements of their program regularly evaluated, 17% did not regularly conduct walkthrough inspections with safety checklists, 43% did not keep records for an equipment maintenance plan, 55% did not incorporate training in hazard recognition, 0% had point of operation guards in place on all equipment, and 78% permitted students to participate in laboratory activities prior to earning 100% on the corresponding safety tests. More recently, Love and Roy's [5] research, which focused more broadly on safety across various TEE and CTE courses, revealed similarly concerning results. For example, among the 718 participating teachers, only 49% had safety data sheets (SDS) readily accessible and only 83% had ANSI/ISEA Z87.1 D3-rated safety glasses with side shields available for every student during laboratory activities [5]. Statistical analyses from that dataset revealed that TEE and CTE classes with enrollments exceeding 24 students were 48% more likely to have had an accident occur within the past five years [32]. Additionally, it was discovered that educators who had comprehensive safety training (a combination of training during their undergraduate or graduate studies, training upon initial hiring at their school district, and a safety training update within the past five years) were 49% less likely to have had an accident occur in the CTE and TEE courses they taught [32]. Other studies have found that safety training, with connections to industry training methods, can enhance educators' self-efficacy for teaching safety concepts and expected safety outcomes of students as a result of their instruction [21]. Furthermore, in a follow-up study focused on the middle Atlantic (mid-Atlantic) region teachers from Love and Roy's national study [32], Love [44] discovered that equipment and machinery were involved in significantly more accidents in the mid-Atlantic region than other regions, and that manufacturing and construction courses had a significantly higher number of accident occurrences compared to other courses. This led Love to suggest that further analyses were needed to examine the differences related to accident occurrences and items involved in accidents among different TEE and CTE courses.

2.7. Rationale and Research Questions

The literature indicates that there is a connection between safety habits and practices that students learn in secondary education TEE and CTE programs, and the transfer of those skills to the workplace and post-secondary programs [9]. The literature has also documented the benefits of safety training for students and instructors; however, the literature lacks research on safety factors associated specifically with accidents in secondary-level construction courses. Therefore, the researchers conducted this study to investigate how safety factors related to construction courses differed from other TEE courses taught in the U.S. Furthermore, one of the main goals of this study was to identify safety factors that were significantly associated with accident occurrences in construction courses to help proactively address these issues and provide safer construction education experiences. To address the gaps identified in the literature and recommendations from prior studies [44], the following research questions (RQ) were developed to guide this study:

RQ1: What are the demographics and characteristics of educators teaching construction courses in the U.S.?

RQ2: Is there a significant difference between the number of accident occurrences in construction courses compared to other TEE courses?

RQ3: What safety factors are significantly associated with accident occurrences in secondary education construction courses?

RQ4: Is there a significant difference between the items involved in secondary education construction course accidents compared to those involved in accidents from other TEE courses?

3. Methods

Data collected from the Technology and Engineering Education—Facilities and Safety Survey (TEE-FASS) [5] were analyzed in this study. The TEE-FASS has a series of demographic and Likert-scale items that ask participating teachers about their demographics, experience, teaching conditions, facility characteristics, safety training, safety practices, and accidents. Since the instrument included a large volume of questions, and it required participating teachers to recall information from previous years (e.g., number of accidents that occurred within the past five years), the instrument was designed to collect mostly nominal and ordinal data to make it more user-friendly. The TEE-FASS used the following definitions to classify minor and major accidents, and these definitions were provided to participants within the survey immediately before the questions pertaining to accident occurrences:

A minor accident encompasses water or chemical spills, slipping on dusty floors, broken glass, excessive fumes, small fires, projectiles, or other accidents during course activities that either resulted in no injuries or required minor medical attention such as Band-Aids, minor first aid, or a visit to the school nurse. Major accidents resulted in an injury during course activities that required major medical attention with a visit to a doctor or hospital (stitches, etc.). [32] (p. 5).

To advertise the survey to key stakeholders and encourage voluntary participation, the link to the TEE-FASS was shared by national and state T&E educator and CTE professional associations. This resulted in 718 responses from TEE and CTE teachers in 42 U.S. states. Additional information about the reliability and validity measures of the TEE-FASS are described in detail by Love et al. [32], and the full instrument was published by Love and Roy [5].

3.1. Data Analyses

Descriptive statistics were used in RQ1 to examine the demographics and characteristics reported by participants. For RQ2, descriptive statistics and Mann–Whitney U tests were utilized. In the TEE-FASS, participants were asked to report accident rates as ordinal responses (e.g., How many accidents occurred within the past five years? Response choices: 0, 1–5, 6–10, 11–15, or >15). A table was created to assist with displaying the occurrence of accidents reported according to the courses. Next, Mann–Whitney U analyses were conducted to test for significant differences between two independent samples with ordinal (accident occurrence categories) and nominal (construction course or other TEE course) data [45].

For RQ3, the researchers used an exploratory correlational analysis method that had been implemented in previous safety studies [31,32,44,46]. This method allowed the researchers to estimate the independent associations between various safety factors reported in the TEE-FASS and the occurrence of accidents over a five-year span. Associations were estimated as polychoric correlation coefficients. The polychoric correlation analysis, which is an alternative to the Pearson r, is used when data is organized in an ordinal manner, yet the variables represent a continuous measure (i.e., accident occurrence categories) [47]. Both the p-value for the likelihood ratio test and the polychoric correlation coefficient were reported for each safety factor. Lastly, Mann–Whitney U tests were again used to investigate differences between items involved in accidents within construction courses compared to items involved in accidents within other TEE courses.

3.2. Participants

Among the 718 TEE-FASS responses, 119 TEE or CTE teachers from 25 different states in the U.S. reported that construction courses were among the top three foci of the classes they taught. Their responses were specific to construction courses. CAD and architectural design courses were a separate survey option, and the researchers did not include those as part of the sample of 119 construction teachers. The demographics and characteristics of these 119 participants compared to the full sample and previous studies are discussed in more detail in RQ1. Furthermore, additional demographic information about the national sample was reported by Love and Roy [5]. The full list of non-construction courses (pre-engineering, electronics, woods and metals materials processing, etc.) taught by participants in the full sample were reported also by Love and Roy [5].

4. Results

4.1. RQ1: Demographics and Characteristics

The first research question examined the demographics and characteristics of educators teaching secondary-level construction courses in the U.S. Descriptive statistics revealed that most participating TEE and CTE teachers in the U.S. identified as white and male. Upon closer examination, there were a higher percentage of white (92%) and male (87%) construction teachers compared to participants teaching other TEE courses. However, when compared to the 2008 national Schools and Staffing Survey (SASS) [48] data for construction, architecture, and engineering technologies teachers, there were more females teaching construction courses in the present study. Most construction teachers in this study had state teaching certification in TEE (87%) and/or a CTE area (9%), which was greater than the percentage of participants teaching other TEE courses (76%) and those who participated in the SASS (73%). In regard to grade level taught, participants teaching construction courses were similar to other TEE teachers, but there was a slightly higher percentage teaching at the high school level (60%). When it came to teaching experience there was a lower percentage of educators within their first nine years of teaching (23%) compared to educators teaching other TEE courses (31%) and those who participated in the SASS (47%). Moreover, 57% of the educators teaching construction courses in this study had been teaching for over 15 years in comparison to only 37% who reported belonging to this category in the SASS (Table 1).

Table 1. Participant Demographics and Characteristics.

Demographic or Characteristic	Const. * n = 119	Other * n = 599	Full Sample * n = 718	SASS ^ n = 9400
Gender				
Male	87%	71%	74%	96%
Female	13%	29%	26%	4%
Ethnicity				
White	92%	90%	90%	89%
Hispanic	2%	1%	1%	NR
Black	4%	5%	5%	9%
Two or more races	1%	3%	3%	NR
State Teaching Certification Area				
TEE	87%	76%	78%	73% #
CTE area	9%	8%	8%	
Science	0%	5%	5%	NR
Alternative Certification	3%	3%	3%	43%
Grade Level Taught				
6–8	25%	30%	29%	NR
9–12	60%	54%	55%	100%
6–12	13%	11%	11%	NR
Years of secondary-level teaching experience				
0–4 years	7%	10%	10%	19%
4–9 years	16%	21%	20%	28%
10–14 years	20%	20%	20%	17%
≥15 years	57%	49%	50%	37%

Note. TEE = technology and engineering education; CTE = career and technical education; Const. = construction teachers; Other = educators of non-construction TEE courses; Full sample = national sample; * = Source: Love and Roy [5]; ^ = Schools and Staffing Survey (SASS) [48] results for construction, architecture, and engineering technologies teachers; # = had full state certification and demonstrated competency in the construction, architecture, and engineering technologies subject area [48]; NR = not reported.

Within RQ1 we also examined the educational background of construction teachers versus other TEE teachers. Participants were asked to report all degrees earned, not just their highest degree. This revealed a slightly higher percentage of construction teachers had an associate degree in an industry area or another education field (e.g., general education courses for state certification). In regard to the bachelor's degree area, construction teachers were similar to other TEE teachers, but a slightly higher percentage of construction teachers had a graduate certificate in an industry area (3%). Interestingly, a higher percentage of construction teachers earned their master's in TEE (21%) than participants teaching other TEE courses (16%). However, among those teaching other TEE courses there was a higher percentage of participants who earned their master's in other educational areas (e.g., administration) (Table 2).

Table 2. Educational Background of Participants.

Degree and Course	Degree Area					
	IA	TEE	STEM	Other Ed	Eng	Industry
Associate Degree						
Const.	2%	1%	0%	6%	3%	8%
Other	2%	2%	0.3%	3%	5%	5%
Full sample	2%	1%	0.3%	3%	5%	6%
Bachelor's Degree						
Const.	0%	29%	2%	8%	8%	3%
Other	0.2%	28%	3%	8%	7%	2%
Full sample	0.1%	28%	3%	8%	7%	2%
Graduate Certificate						
Const.	5%	3%	2%	5%	0%	3%
Other	1%	4%	3%	4%	0.5%	1%
Full sample	2%	4%	3%	4%	0.4%	1%
Master's Degree						
Const.	4%	21%	4%	26%	0%	3%
Other	6%	16%	5%	29%	3%	2%
Full sample	6%	17%	5%	29%	2%	2%

Note. IA = industrial arts; TEE = technology and engineering education; STEM = science, technology, engineering, and mathematics education; Other Ed = education field not related to the others listed in the table; Eng = Engineering field (non-education-related); Const. = construction teachers; Other = educators of non-construction TEE courses; Full sample = national sample. Const. n = 119; Other n = 599; Full sample n = 718. Source: Love and Roy [5].

Next, we examined participants' experiences relative to safety training. A higher percentage of participating construction teachers received safety training within their undergraduate (66%) or graduate (31%) coursework compared to other TEE teachers. Overall, a low percentage of construction and other TEE teachers received safety training from their school district upon initial hire and a safety training update from their school district within the past five years. When examining the content of the safety training provided by school districts, it was found that participating construction teachers received less training on all topics except for federal or state OSHA standards. In regard to those who received their safety training update from a source outside of their school district, a much higher percentage of construction teachers received training from a university (21%) or national safety company (13%), but a higher percentage of other TEE teachers received training from their state's department of education (8%) or a manufacturer/curriculum provider (6%). When examining what was deemed as a comprehensive training experience [32], a slightly higher percentage of construction teachers (71%) had comprehensive safety training (Table 3).

Lastly, for RQ1 we examined average class enrollment size (occupancy load) of construction classes. Fifty-two percent of construction teachers reported having class sizes above 20 students which was higher than the full sample (47%), but lower than other TEE courses (56%). Construction courses had a higher percentage (37%) of courses in the 16–20 students range compared to other TEE courses (32%) and the full sample (33%). When comparing the construction courses from this study to those from the SASS [48], the data indicates that average class sizes have increased over the past 15 years. The SASS reported that only 29% of construction, architecture, and engineering technologies teachers had average enrollments above 20 students compared to 52% of the participants in the present study (Table 4).

Table 3. Safety Training Experiences of Participants.

Safety Training Experience	Const. n = 119	Other n = 599	Full Sample n = 718
Training in Undergraduate Coursework	66%	61%	62%
Training in Graduate Coursework	31%	28%	28%
Training Upon Initial Hire	31%	32%	32%
Training Update from School District	55%	56%	56%
School District Training Update Covered:			
Federal or state OSHA standards	74%	73%	73%
School district Hazard Communication plan	60%	65%	64%
Reading globally harmonized system (GHS) labels	48%	54%	53%
Chemical/Paint/Solvent Storage and Disposal	59%	62%	61%
Safety data sheets (SDS)	66%	73%	72%
Managing unsafe student behaviors	55%	59%	58%
First-aid procedures	71%	78%	77%
Training Update from an Ind. Source	20%	18%	18%
Source of Ind. Training Update:			
Local source	29%	32%	32%
State educator association	17%	18%	18%
State department of education	4%	8%	7%
National educator association	0%	5%	4%
University	21%	10%	12%
OSHA	17%	18%	18%
Manufacturer or curriculum provider	0%	6%	5%
National safety company	13%	4%	5%
Comprehensive safety training	71%	66%	67%

Note. Ind. = Independent; Const. = construction teachers; Other = educators of non-construction TEE courses. Source: Love and Roy [5].

Table 4. Average Course Enrollment Size.

Number of Students	Const. * n = 119	Other * n = 599	Full Sample * n = 718	SASS ˆ n = 9400
$\leq$15	12%	12%	12%	44%
16–20	37%	32%	33%	26%
21–24	23%	25%	25%	22% #
25–30	19%	23%	22%	22% #
>30	10%	8%	8%	7%

Note. Const. = construction teachers; Other = educators of non-construction TEE courses; Full sample = national sample; * = Source: Love and Roy [5]; ˆ = Source: Schools and Staffing Survey (SASS) [48] results for construction, architecture, and engineering technologies teachers. # = The SASS reported results as a category of 21–30 students.

4.2. RQ2: Accident Occurrences

The second research question investigated if there was a significant difference between the number of minor and major accidents that occurred over a five-year span in secondary education construction courses and other TEE courses in the U.S. Descriptive statistics were first used to examine the percentage of accidents reported within each range of accident occurrences. This showed that a much higher percentage (17%) of teachers in other TEE courses reported having no minor accident occurrences in their courses compared to construction courses (8%). Moreover, there were a higher percentage of minor accidents in construction courses for the 6–10, 11–15, and >15 accident occurrence categories. There was also a slightly higher percentage (33%) of 1–5 major accident occurrences for construction courses. These percentages led the researchers to hypothesize that there was a significantly greater occurrence of minor accidents in construction courses compared to other TEE courses (Table 5).

Table 5. Accident Occurrences During a Five-Year Span.

Accident Type and Course	Number of Accidents				
	0 (%)	1–5 (%)	6–10 (%)	11–15 (%)	>15 (%)
Minor Accidents					
Construction	8	46	24	13	9
Other Courses	17	49	17	10	7
Major Accidents					
Construction	66	33	0.8	0	0
Other Courses	69	31	0.2	0	0

Note. Construction n = 119, Other courses n = 599.

To examine if there was a significant difference between the number of accidents that occurred in construction courses and other TEE courses in the U.S., Mann–Whitney U tests were conducted. These analyses revealed a significantly greater occurrence of minor accidents in construction courses over a five-year period ($p = 0.003$), but there was not a significant difference in the number of major accident occurrences (Table 6).

Table 6. Mann-Whitney U tests for Accident Occurrences During a Five-Year Span.

Accident Type and Course	Median	Mean Rank	U	z	p
Minor Accidents					
Construction	1	407.89	29,881.5	−2.983	0.003 *
Other Courses	1	349.89			
Major Accidents					
Construction	0	368.28	34,595.5	−0.629	0.529
Other Courses	0	357.76			

Note. Construction n = 119, Other courses n = 599, * = $p < 0.05$.

4.3. RQ3: Factors Associated with Accident Occurrences

After examining the differences between accident occurrences in construction courses and other TEE courses in the U.S., the third research question utilized polychoric correlation analyses to investigate what safety factors were significantly associated with accident occurrences in construction courses during the prior five years. The polychoric correlation analyses revealed the direction of the correlations, which the researchers reported in Table 7 as risk factors (positive correlation) or protective factors (negative correlation). The polychoric correlation analyses suggest that as a risk factor was present or increased, the number of reported accidents also increased. Conversely, as protective factors were present or increased, the number of reported accidents decreased. Seven risk factors and 13 protective factors were found to be significantly associated with reported accident occurrences. To assist with the interpretation of these results, the researchers organized these factors into categories consisting of: educational/credential factors, facilities and equipment factors, standard operating safety procedure factors, administrative factors, and safety training factors. The risk factor category with the most items was facilities and equipment, whereas the largest protective factor category was safety training.

Alternative certification was found to be a risk factor; however, possessing a bachelor's degree in an engineering field was found to be a protective factor. Having a laboratory in the instructional area or connected to the instructional area was significantly associated with an increase in accident occurrences. As students spent more time doing hands-on activities, the number or minor accidents also increased significantly. Finishing rooms were found to be significantly associated with minor and major accident occurrences, and student access to storage areas were significantly correlated with minor accidents. Allowing students to independently use a table saw in the laboratory demonstrated marginal significance with major accident occurrences. Additionally, courses that had average enrollments exceeding

20 students were linked to major accident occurrences with marginal significance. In regard to protective factors, having lockable storage cabinets, a sink in the instructional space, and a fume extractor for soldering activities were all linked to minor accident occurrences with moderate significance. When school districts had SDS on file, this was found to be significantly associated with fewer minor accident occurrences. Among various safety training factors, school district training on safer classroom management strategies was the only training factor found to be significantly associated with reduced minor and major accident occurrences. A number of other safety training factors were found to be significantly associated with decreased occurrences of either minor or major accidents. Since the TEE-FAS included a large volume of questions about safety factors, only those which were found to be statistically significant were reported in Table 7.

Table 7. Polychoric Correlations of Safety Factors Associated with Minor and Major Accidents During a Five-Year Span in Secondary Education Construction Courses.

Significant Safety Risk Factors	Minor Accidents		Major Accidents	
	ρ	p	ρ	p
Educational/credential factors				
Alternative certification	0.64	*	0.66	*
Facilities and equipment factors				
Lab in/connected to facility	0.48	*	0.46	*
Separate finishing room	0.39	**	0.29	*
Table saw used in lab	0.21		0.38	~
Standard operating safety procedure factors				
Percentage of class time doing hands-on activities	0.26	*	0.06	
Student access to storage areas	0.29	**	0.14	
Administrative factors				
Average class size over 20 students	0.09		0.29	~
Significant Safety Protective Factors	Minor Accidents		Major Accidents	
	ρ	p	ρ	p
Educational/credential factors				
Bachelor's degree in engineering	−0.40	*	−0.38	**
Facilities and equipment factors				
Lockable storage cabinet(s)	−0.23	~	−0.07	
Sink in classroom/lab	−0.25	~	−0.19	
Soldering fume extractor	−0.38	~	−0.15	
Administrative factors				
Safety data sheets (SDS) on file with school district	−0.26	*	−0.03	
Safety training factors				
Safety training in graduate courses	−0.30	*	−0.22	
District training on OSHA standards	−0.40	*	−0.23	
District training on classroom management	−0.42	**	−0.49	**
District training on Hazard Communication plan	−0.17		−0.34	~
District training on SDS	−0.23		−0.42	*
District training on globally harmonized system (GHS)	−0.10		−0.53	**
District training on chemical storage and disposal	−0.16		−0.46	**
District training on First-aid	−0.26		−0.49	**

Note. ** = $p < 0.01$, * = $p < 0.05$, ~ = $p < 0.10$.

4.4. RQ4: Items Involved with Accidents

Although RQ3 examined the associations between safety factors and accident occurrences in construction courses, it did not investigate specific tools, equipment, or other items that were reported as being involved with accidents. Therefore, in RQ4 the researchers examined if there was a significant difference between items involved with accidents in construction courses compared to those involved with accidents in other TEE courses. Similar to RQ2, Mann–Whitney U tests were conducted, and these analyses revealed that hand and power tools (screwdrivers, chisels, utility knives, cordless drills, nail guns, etc.) was the only item that significantly differed ($p = 0.002$) between construction courses and other TEE courses in the U.S. (Table 8).

Table 8. Mann–Whitney U tests for Items Involved in Accidents During a Five-Year Span.

Item and Course	Involved n (%)	Median	Mean Rank	U	z	p
Hot Glue Guns						
Construction	38 (32)	0	339.14	33,217.5	−1.397	0.162
Other Courses	232 (39)	0	363.55			
Hand/Power Tools						
Construction	37 (31)	0	396.62	31,223.0	−3.043	0.002 *
Other Courses	112 (19)	0	352.13			
Equipment/Machinery						
Construction	32 (27)	0	374.04	33,910.5	−1.151	0.250
Other Courses	132 (22)	0	356.61			
Automated Equipment						
Construction	4 (3)	0	357.07	35,351.0	−0.411	0.681
Other Courses	25 (4)	0	359.98			
Projectiles						
Construction	20 (17)	0	369.84	34,410.5	−0.992	0.321
Other Courses	80 (13)	0	357.45			
Electrical Short						
Construction	6 (5)	0	355.10	35,117.0	−0.603	0.546
Other Courses	39 (7)	0	360.37			
Fires						
Construction	4 (3)	0	364.57	35,037.5	−1.218	0.223
Other Courses	10 (2)	0	358.49			
Broken Glass						
Construction	40 (4)	0	352.08	34,758.0	−1.017	0.309
Other Courses	5 (7)	0	360.97			
Spills/Splashes						
Construction	22 (19)	0	372.87	34,049.5	−1.253	0.210
Other Courses	84 (14)	0	356.84			

Note. Involved = number of participants who reported this item was involved in an accident within the past five years, Construction n = 119, Other courses n = 599, * = $p < 0.05$.

5. Discussion

The results in this section are discussed according to each research question.

5.1. Demographics and Characteristics

Between 2011 and 2021, the demographics of workers in the construction industry (encompassing various construction fields) changed. The percentage of construction workers 55 or older increased to 22%, the percentage of Hispanic employees increased to 33%, and the percentage of female construction workers increased to 11% [2]. When examining the demographics of the participants in this study, it is clear that the demographics of secondary construction teachers are also shifting. There was a higher percentage of female teachers in this study (13%) compared to the SASS results (4%). Table 1 also reveals changes in years of teaching experience over the past 15 years from the SASS results to the present study. The data indicate that there are fewer construction teachers who are new to the profession. While this may appear as a positive finding related to the retention of experienced and highly qualified construction teachers, it may also forecast potential issues as aging construction educators retire and there is a lack of newer teachers in the pipeline to continue leading these programs. This reflects safety concerns from the literature related to TEE and CTE teacher recruitment, retention, and increases in the number of TEE and CTE teachers who have alternative certification [7,21,22,25]. However, data in Table 1 also reveal that in comparison to the SASS results, the percentage of construction teachers that have earned full state certification has increased. This may be reflective of the decrease in newer teachers since most states have a specific amount of time in which educators must complete coursework transitioning them from alternative certification to a full certification. The increase in certified instructors teaching construction is encouraging to see, as the courses counted toward state certification should be covering important topics such as safety and classroom management practices.

One other noticeable finding from Table 1 is that although the percentage of construction teachers consisted of mostly White participants, there were slightly more Hispanic participants (2%) teaching construction courses compared to other TEE courses (1%). Recruiting, hiring, and retaining highly qualified educators from diverse backgrounds to teach construction courses at the secondary level is very important given the increasing diversity reflected in the construction workforce data. Teacher preparation programs and school districts should seek to hire qualified teachers and support staff from diverse backgrounds to assist and recruit students from diverse backgrounds into secondary construction programs. This can potentially help recruit and prepare more students from diverse backgrounds for rewarding careers and post-secondary opportunities in construction.

While the SASS [48] did not report on specific degree areas, it did indicate that 26% of construction, architecture, and engineering technologies teachers had an associate degree or certificate, 32% had a bachelor's degree, and 37% had a master's degree as their highest educational attainment. When examining the data from this study it is difficult to compare to the SASS because participants could select multiple fields for the same degree level (e.g., if a participant had two associate degrees in different fields). Table 2 does provide insight about the type of fields in which construction educators were prepared. There was a larger percentage of construction teachers with an associate degree in an industry area (8%) compared to other TEE teachers (5%). In regard to bachelor's degrees, construction teachers were similar to other TEE teachers with 29% possessing a bachelor's degree in TEE. A higher percentage of construction teachers had a graduate certificate in industrial arts (5%) or an industry area (3%). This finding aligns with the results from Table 1, as the field of industrial arts officially changed its name to technology education in 1985. Given the high percentage of construction teachers with over 15 years of teaching experience, this finding reflects the preparation experiences of those teachers. It is also worth noting that while concepts focused on construction technologies were part of previous technology education standards [49] and the built environment is part of current TEE standards [6], teaching about construction had a more prominent role in industrial arts curricula and teacher preparation programs. For example, the Jackson's Mill Industrial Arts Curriculum Theory [50], which guided the field for many years, identified construction as one of the four subsystems of human technical endeavor in which students should have learning experiences and that industrial arts/technology education teachers should be prepared to facilitate student learning. Experienced teachers who had an industrial arts preparation experience may have received more preparation in teaching construction concepts than newer teachers who completed a program aligned with more recent standards focused on applying the design process within various T&E contexts. Moreover, a higher percentage (21%) of construction teachers completed a master's degree in TEE, but fewer received their master's degree in other educational fields. This may be reflective of the higher percentage of construction educators who earned degrees in industry areas and wanted a master's degree related to teaching such concepts.

5.1.1. Safety Training Experiences

The SASS [48] reported that 72% of construction, architecture, and engineering technologies teachers received PD related to their subject area during the 2007–2008 academic year. Among those teachers, 21% completed eight or less hours of PD, 27% completed nine to 16 h, 20% completed 17–32 h, and 33% completed 33 or more hours of PD that year. While the SASS data on PD completion are not specific to safety training, this data provide an overview of the time teachers spent receiving PD on various topics during the year. Within the current study, a slightly higher percentage of construction teachers received some form of safety training in their undergraduate (66%) and graduate (31%) coursework compared to other TEE teachers. However, this percentage is still very low considering safety should be a core component of all undergraduate and graduate programs in TEE and CTE fields [32]. The lack of safety training at this level is consistent with findings from safety training studies [21]. When examining the percentage of construction teachers who received safety training from their school district upon initial hiring and a safety training update within the past five years, these results are similar to other TEE teachers, but also very low. OSHA's Hazard Communication Standard (29 CFR 1910.1200), in addition to specific OSHA legal standards, require employers (school districts) to train employees (teachers) upon initial hiring, when there are changes in work assignments, and when there are changes in safety plans and workplace safety hazards or risks [5,32,51]. In regard to content included in the safety training updates provided by school districts or independent sources, Table 3 demonstrates there was a limited focus on various areas that OSHA indicates must be part of safety training offered by an employer [5,21,51]. Additional information about the importance of these training areas is discussed in a later section focused the significant safety training factors. When examining the independent source who provided safety training updates, universities and national safety companies were more prevalent among construction teachers compared to other TEE teachers. This may indicate that school districts are using companies with specialized expertise in construction safety to deliver more focused training for their teachers. It could also suggest that universities with construction and OSH programs have unique opportunities to offer specialized safety training for construction, CTE, and TEE teachers that allow for safer preparation of prospective students and building relationships with local school districts for recruitment purposes. Lastly, one positive finding in regard to training was that a higher percentage of construction teachers (71%) had completed a comprehensive safety training experience compared to other TEE teachers (66%). However, teacher preparation programs and school districts should collaborate to increase the number of teachers who have a comprehensive training experience as this was found in previous studies to decrease the odds of an accident by 49% [32].

5.1.2. Average Class Enrollment Size

When examining average course enrollment size, it was encouraging to see that a lower percentage of construction courses had 21 or more students compared to other TEE courses; however, over half (52%) of the participating construction teachers reported having average enrollments surpassing 20 students. Another alarming finding is that when compared to the SASS data, it is apparent that average class sizes in construction courses have increased over the past 15 years. In the 2008 SASS results, 70% of construction teachers had average class sizes of 20 or fewer students. The current study reveals that this percentage has dropped to 49%. The increase in average course enrollment size may be reflective of challenges related to finding highly qualified teachers who are certified to teach construction courses. As construction teachers retire or pursue other career opportunities, school districts may be tempted to place more students into the construction courses taught by the remaining educator(s). While this may seem like a quick solution and save funding, it limits the supervision and assistance that a teacher can provide to students, increasing the chance of an accident as discussed in a later section on safety factors associated with accident occurrences.

5.2. Accident Occurrences

Analyses in RQ2 found that 92% of participants reported having a minor accident occurrence, while 34% had a major accident occurrence in their construction courses over the past five years. This is a stark increase from Nichols's [43] study which found that 42% of construction teachers had accident occurrences involving students, and eight percent had a student sustain a serious injury over a five-year span. The statistical analyses in the current study revealed there were a significantly greater number of minor accident occurrences in construction courses taught in the U.S. when compared to other TEE courses; however, the analyses also revealed there was no significant difference in the number of major accident occurrences. This aligns with findings from a previous study, which found that in the mid-Atlantic region of the U.S. there were significantly more accident occurrences in construction and manufacturing courses than other TEE courses [44]. These findings also reflect the inherently hazardous nature of the field of construction, which data indicates is one of the most hazardous industries in the U.S. [4]. Furthermore, these findings reiterate the importance of recruiting, hiring, and retaining highly qualified educators to teach construction courses at the secondary level. It is critical that school districts support their construction teachers through ongoing safety training, adequate funding for safety needs, and other resources, such as those described in the next section, to reduce the odds of an accident occurring.

5.3. Safety Factors Associated with Accident Occurrences

When examining the significant safety factors reported in Table 7, it is important to consider the exploratory nature of the polychoric correlation analyses that were conducted in this study. Many of the safety factors listed in Table 7 are mandated by state OSH or federal OSHA plans (e.g., OSHA's Hazard Communication Standard involving SDS—General Industry 29 CFR 1910.1200 and Construction 29 CFR 1926.59) and are also legally required under better professional safety practices. One observation from Table 7 is that it is interesting to see that some safety factors were significantly associated with minor accident occurrences but not with major accident occurrences, or vice versa. The polychoric correlation tests suggest that as the protective factors are implemented or increased, or as the risk factors are reduced or eliminated, theoretically, the chance of an accident should decrease. It is important to remember that this list does not encompass all legal and safety practices that may be required (e.g., requiring students to earn a 100% on all safety tests before using any potentially hazardous tools or equipment); rather, these are the safety factors that were found to be significantly associated with accident occurrences in the construction courses taught by participants in this study. It is also apparent that some safety factors in Table 7 had a smaller correlation coefficient than other factors. While this does not suggest that specific safety factors are more hazardous than others, it merely highlights safety factors that may need additional attention to provide safer construction instruction at the secondary level. For more details about the safety factors that were significantly correlated with accident occurrences in other studies and how they relate to state and/or federal OSH regulations, please see Love and Roy [5] and Love et al. [32]. The safety factors found to be significant in this study are discussed in detail below.

5.3.1. Educational and Credential Factors

While previous studies expressed anecdotal concerns regarding the increase in the percentage of educators who were teaching in TEE and CTE laboratory settings with alternative certification, they did not present any data documenting an association between the types of certification and accident occurrences [7,21,23]. Analyses from the present study revealed that having alternative certification was significantly correlated with increases in minor and major accident occurrences. School districts that hire TEE and CTE teachers with alternative certification should ensure that they receive the training and support required as newer educators. This includes support to complete coursework specific to TEE or CTE teaching and help in the transition from alternative to full state certification. It is plausible to hypothesize that coursework specific to TEE or CTE teaching is arguably more valuable for these educators compared to generic education courses, because TEE and CTE courses should cover unique topics, such as safety and facilities management practices. The correlation analyses also discovered that having a bachelor's degree in an engineering (non-education) field was significantly associated with a decrease in accident occurrences. Those who had a degree in an engineering field should have completed coursework (lectures and labs) which presented various safety considerations and practices as required for post-secondary engineering program accreditation. Knowledge of authentic safety considerations and applications could benefit teachers as they transition from industry to teaching. This finding mirrors results from another TEE and CTE teacher safety study which found that having a graduate certificate in an engineering field was a marginally significant safety protective factor [31].

5.3.2. Facilities and Equipment Factors

Construction teachers who had a facility with a laboratory or had access to a laboratory connected directly to their facility reported significantly higher occurrences of minor and major accidents. This is not a surprising finding as more potentially hazardous and higher resulting risk activities should be conducted in laboratories that have the proper engineering controls as opposed to general classroom spaces. Various other studies also found laboratories in or connected to facilities to be a significant risk factor [31,32,44,46]. Like previous studies that examined safety issues in the mid-Atlantic [44] and northeast [31] regions of the U.S., having a separate finishing room was a significant risk factor for both minor and major accident occurrences in this study. As described by Love and Roy [5], finishing rooms can offer protective safety benefits such as externally vented paint booths which can help isolate and limit particulate exposure in a facility. However, these spaces can also create additional areas for instructors to supervise and host an array of additional chemical and physical hazards that need to be addressed. Similar to finishing rooms and storage areas, lockable storage cabinets also require directly supervised access due to the potentially hazardous items in these areas. OSHA and the National Fire Protection Association (NFPA) have strict legal requirements for separate finishing rooms, chemical storage areas, and lockable cabinets used to store flammables. Like previous studies [32,44], lockable storage cabinets were found to be a marginally significant safety protective factor. When used appropriately, lockable chemical storage and tool storage cabinets can help to limit theft and unapproved use of hazardous materials, reducing the chance of an accident and limiting the liability of the instructor and the school district [5,52].

Having access to a sink in the construction course facility was a marginally significant protective factor, as previous TEE and CTE studies discovered [31,32,44]. Sinks and access to potable water are important in laboratory course areas because students should be washing their hands with soap and water after handling potentially hazardous materials (e.g., treated lumber) and before leaving class. Access to clean running water may also be required in certain emergency situations. Better professional safety practice suggests each lab should have one sink per every four students [53]. Additionally, the use of soldering fume extractors was found to be a marginally significant protective factor. This could apply to various soldering-related applications, such as brazing and other processes used in plumbing construction applications. More information about soldering and other ventilation requirements are described by Love and Roy [5].

Lastly, in regard to facilities and equipment safety factors, the use of a table saw in a construction course was found to be a marginally significant ($p < 0.10$) safety risk factor associated with major accidents. This aligns with previous studies which found the use of a table saw to be a significant safety risk factor in various TEE and CTE courses [31,32,46]. However, it should be noted that in two previous studies [32,44], it was also discovered that teachers who had a table saw with the patented SawStop safety technology had significantly fewer accident occurrences than instructors who had a table saw without the SawStop safety feature. As suggested in previous studies [5,32], due to the high risk of accidents when using a table saw and the horrific nature of table saw accidents described in the case law, better professional practices and legal safety standards indicate that school districts should invest in a SawStop table saw or SawStop jobsite/contractor saw to significantly reduce the chance and severity of an accident. If a construction teacher has multiple table saws that they use in their courses, better professional practices and legal safety standards suggest students should only be allowed to use the SawStop model.

5.3.3. Administrative Factors

One administrative factor significantly associated with decreased minor accident occurrences was the school district keeping copies of SDS on file. This was also a significant protective factor in Love and Roy's study of pre-engineering teachers [46]. As explained by Love and Roy [5], under OSHA's Hazard Communication Standard (29 CFR 1910.1200), employers (school districts) must ensure that SDS for all hazardous chemicals and materials are readily accessible to employees (teachers), and should provide employees with effective information and training in this area. Moreover, copies of all SDS should be kept on file by the TEE or CTE department, the school nurse, the district facilities office/safety director, and the local fire marshal.

The other administrative factor that was found to be correlated with increased accident occurrences was average class enrollment size (occupancy load) exceeding 20 students. This safety risk factor was not significantly associated with minor accident occurrences but a marginal significance ($p < 0.10$) associated with increased major accident occurrences was indicated. This corroborates decades of concerns, research findings, NFPA 101 Life Safety Code Occupancy Load standards, and recommendations regarding safer occupancy load limits in TEE and CTE laboratory courses [31,32,44,46,54]. When examining the greatest causes of accidents reported by the full national sample [5], teachers selected overcrowding as the second greatest cause for accident occurrences. Furthermore, numerous studies have found accident occurrences to significantly increase when occupancy loads exceed 24 students per one instructor [31,32,44,46,54–56]. However, in this study specific to secondary-level construction courses, major accident occurrences significantly increased when average course occupancy loads exceeded 20 students. This may be reflective of the unique and increased hazards that are present in construction classes (heavy concrete masonry units, ladders, long stock materials, electrical hazards, heavy equipment and machinery, etc.) compared to other TEE courses (pre-engineering, electronics, etc.). As discussed in RQ2, construction courses had a significantly greater occurrence of accidents and the construction industry is one of the most dangerous [4]. Recognizing these increased hazards related to construction, states like the Commonwealth of Virginia have passed legislation that limits the enrollment of construction classes to no more than 20 students [57]. However, when looking at the average class enrollments reported in Table 4, it is apparent that 52% and 29% of construction teachers are being tasked with teaching dangerous class sizes exceeding 20 students and 24 students, respectively. As Love and Roy [5] describe in detail, the NFPA 101 Life Safety Code specifies that in educational shops, laboratories, and vocational rooms, a minimum of 50 net square feet per occupant is required [58] (pp. 101–183). However, school administrators, school counselors, and others making decisions about class sizes must remember that although a room may have enough net square footage to host more than 20 occupants, the materials and equipment that will be used in construction courses pose potentially greater risks than other laboratory-based classes. Increasing class enrollment may also be in violation of the NFPA 101 Life Safety Code Occupancy Load legal safety standard and/or better professional safety practices. While the literature suggests that no TEE or CTE teacher have more than 24 students enrolled in their class, the data from this study indicate that construction courses should be limited to no more than 20 students per one certified and safety trained instructor. This represents safer, data-informed professional practices based on the information presented in this article.

5.3.4. Standard Operating Safety Procedure Factors

As the percentage of time spent doing hands-on activities in construction courses increased, so did minor accident occurrences. This finding is not rare, as it has been significantly associated with increased accident occurrences in other safety studies [31,32,44,46]. It is plausible to expect that the more time students are engaged in potentially hazardous hands-on construction learning activities, the more opportunities there are for an accident to occur. However, this does not suggest that authentic hands-on learning opportunities should be reduced or eliminated given the valuable role they play in preparing TEE and CTE students to be career- and college-ready [5,59]. Rather, this finding reinforces the importance of conducting a potential hazards analysis, resulting health and safety risk assessment [5,52], and implementing the safety protective factors described in this article to help reduce the odds of an accident while also maintaining a focus on safer teaching and hands-on learning. Another significant safety factor associated with increased accident occurrences in this category was student access to storage areas. Like finishing rooms, which were previously discussed, storage areas can be difficult for instructors to supervise and provide additional areas for mischief as well as access to potential chemical and physical hazards [32]. Students should not be allowed in storage areas without instructor supervision, and these areas should remain locked when a safety trained instructor is not present. Not only does this help to limit potential accidents from falling items and other safety issues, but it also limits the theft of potentially hazardous items which could ultimately lead to an instructor being found negligent or reckless for not securing this area or directly supervising students while in the storage area [5,52]. Access to storage areas was also found to be a significant risk factor in a previous study examining pre-engineering courses [46].

5.3.5. Safety Training Factors

Numerous safety training factors were found to be significantly associated with reduced minor or major accident occurrences. These results (Table 7) are similar to findings from previous studies [31,32], which discovered that while various safety risk factors increased the odds of an accident occurrence (e.g., class enrollment size), safety training factors helped to reduce the strength of association between certain risk factors and accident occurrences. While the present study did not examine the interaction effect between safety risk and protective factors, it did reveal that a variety of safety training experiences (most of which were provided by the teachers' school districts) were associated with reduced accident occurrences. OSHA's Hazard Communication Standard for General Industry (29 CFR 1910.1200) and Construction (29 CFR 1926.59) requires school districts to train teachers of construction courses on specific content and topics such as those listed in Table 7. Therefore, it would be better professional practice for districts to provide initial and ongoing training about these safety factors that directly relate to safer teaching of construction concepts. This can also benefit the students as teachers can then share their knowledge of industry OSHA standards with students to help them develop industry-accepted safety practices. Furthermore, these results align with Love et al.'s [31] study which identified training related to first-aid information, the school district's hazard communication plan, and safer classroom management strategies as significant safety protective factors in TEE and CTE courses. Love et al. [21] provided detailed recommendations for integrating OSHA-aligned content within safety training for TEE and CTE teachers. The results in this study also align with findings from Love and Roy's [5] national study in which teachers believed that classroom management was one of the top causes for accidents. In the current study, school district training on safer classroom management strategies was significantly associated with the decreased occurrence of minor and major accidents. One area where this study differed from previous research was in regard to safety training that teachers received as part of their post-secondary courses. Love et al. [31] found that undergraduate courses that included some form of safety training were significantly associated with reduced accident occurrences, whereas the current study discovered that graduate level courses involving safety training were significantly associated with fewer accident occurrences.

5.4. Items Involved with Accidents

The fourth research question revealed that, despite the level of risk inherent with large equipment, machinery, automated equipment, and other hazards present in construction courses, the only items that were involved in significantly more accident occurrences in construction classes compared to other TEE courses were hand and power tools. This may be the result of proper guarding (29 CFR 1910 Subpart O and 29 CFR 1926 Subpart I) and other required engineering controls found in OSHA 29 CFR 1910 and 29 CFR 1926 (e.g., PPE) being in place and used correctly to limit accident occurrences resulting from equipment, machinery, automated equipment, electrical equipment, and other items used in construction settings. These results differ from previous studies which found that hot glue guns were involved in significantly more accidents in secondary pre-engineering courses [46] and that equipment and machinery were involved in significantly more accidents in mid-Atlantic TEE and CTE courses [44]. This reinforces an earlier point made about the unique hazards and risks involved in construction courses, which can require specialized safety and pedagogical training to reduce risks and provide safer learning experiences. In light of these findings, safety training for construction teachers and students should emphasize safer practices related to hand and power tools. OSHA general industry standards provide safety information for hand and portable power tools and other hand-held equipment in 29 CFR 1910 Subpart P. Additionally, OSHA construction industry standards provide safety information for hand and power tools in 29 CFR 1926 Subpart I. Moreover, instructors have a legal and ethical obligation to align their instruction with safer practices, including industry standards which students will be expected to follow when they enter the workplace. Construction teachers should demonstrate and enforce safety requirements in their courses that are aligned with the aforementioned OSHA standards. Additional resources to assist with safer training, testing, use, and supervision of construction hand/power tools and equipment are provided by OSHA [60], ITEEA [61], the Power Tool Institute [62], Virginia Tech [63], and others.

5.5. Limitations

Although this study revealed some important findings to help improve safety in secondary-level construction courses, and potentially the safety habits of young workers matriculating from these programs into construction-related fields, there are a few limitations. This study involved a homogenous sample consisting of mainly white and male participants. However, the demographics in this study are similar to Williams and Ernst's [27] national study examining the demographics of TEE teachers. Another limitation is the voluntary self-reporting aspect of the data collected. It is unknown if participating educators had an increased interest in responding due to safety issues they personally experienced and wanted to report to support positive changes. Other potential limitations related to the data collection include asking participants to report accident occurrences as ordinal data instead of continuous data. While this was done to make the instrument more user-friendly for the respondents, it required the use of polychoric correlation analyses, which have some limitations in comparison to Pearson's correlation analyses. Readers should interpret the polychoric correlation results with caution as they do not indicate direct causation of an accident. These results merely indicate that there is a significant association (either positive or negative) between the specified safety factors and accident occurrences. These data, which address a critical gap in the literature, provide implications for raising awareness of, and improving, the safety factors that were significantly associated with accidents. In theory, the data indicate that this should help to significantly reduce the odds of accident occurrences in secondary-level construction courses.

6. Conclusions

Like previous studies [9,38], the results of this research raise some alarming concerns regarding safety in construction courses taught within U.S. secondary schools. Of notable concern are the dangerous class enrollment sizes of construction courses and the lack of safety training among construction teachers, especially with the rising numbers of alternatively certified TEE and CTE teachers. The results from this study have important implications not only for improving safety and limiting liability in secondary TEE and CTE programs, but also for the safety and liability of construction employers and post-secondary construction programs who will be hiring and/or educating these students. As students matriculate into post-secondary education programs and into the workforce, they will carry with them the safety knowledge and practices they developed during their secondary education experience. Post-secondary construction programs and construction industry partners should consider the results of this study when developing safety training and proactive measures to limit potential hazards that the data revealed are significant issues for incoming students and young workers. This could potentially address gaps in students' safety knowledge and practices, and also reduce the odds of an accident. The findings from this study can help to inform collaborative efforts among post-secondary construction programs, secondary TEE and CTE programs, and construction industry partners to address significant safety gaps. Addressing these gaps can contribute to safer, and subsequently more sustainable practices in the construction industry. Lastly, this research has direct implications for secondary TEE and CTE teachers, administrators, school counselors, school districts, school district safety officers, chemical hygiene officers, and TEE and CTE teacher preparation programs involved in making construction teaching and learning experiences safer.

6.1. Recommendations

The following recommendations for researchers and practitioners involved with construction education opportunities were derived from the data.

6.1.1. For Future Research

This study and other recent safety studies, indicate that significant safety factors can be unique to specific courses and regions of the U.S. [44,46]. Therefore, additional studies are warranted to examine which safety factors are significantly associated with accident rates in other TEE and CTE courses (e.g., woods and metals processing courses). Following the methods utilized in other safety studies, further analyses should be conducted to investigate the interaction effects between the significant risk and protective safety factors identified in this study, and their influence on accident occurrences. Lastly, future research should continue to monitor trends in class enrollment sizes and correlations with accident occurrences.

6.1.2. For Practitioners

Bush and Andrews [38] provided a brief list of recommendations which are still relevant and also apply to this study. Their recommendations included: conducting future research with more construction teachers; developing a place for instructors to share best practices and resources; providing greater support for mandating industry-recognized OSH training for secondary construction students (e.g., OSHA 10 h construction standards training); providing greater critical thinking and problem-solving opportunities for students during OSH training experiences; and collaborating with industry partners/advisory committees to place a greater focus on safety programs for students and young workers. Roy and Love [5] provided examples of pedagogical strategies that can assist with making safety instruction more engaging and increase opportunities for critical thinking. In addition to collaborating with industry partners, school districts should partner with their state department of education, neighboring school districts, secondary construction teachers, school district safety and chemical hygiene officers, and TEE and CTE teacher preparation programs to develop PD that addresses areas of need from the data presented in this study.

With the literature showing increases in the percentage of TEE and CTE teachers who have alternative certification [7,21,23,27,28], and the significant correlation between alternative certification and increased accident occurrences in this study, recruitment and retention of highly qualified construction teachers is critical. This is especially true for recruiting and retaining a more diverse workforce of construction teachers. Contributing to retention is the support that school districts, teacher preparation programs, and professional associations provide for newer and alternatively certified teachers. Ensuring that these teachers complete TEE and CTE specific coursework to obtain their full certification and receive safety training on issues described in this article will be crucial for retention and safer construction instruction. Lastly, administrators, school counselors, and others who make decisions on class enrollment sizes should not place more than 20 students in a construction course with one certified and safety trained educator. Additional support may be required to help provide the appropriate modifications and accommodations for students of all abilities in construction courses. Based on the findings from the statistical analyses in this study, it would be better professional practice to limit the enrollment in secondary construction classes to 20 students per one certified and safety trained educator.

Author Contributions: Conceptualization, T.S.L.; methodology, T.S.L.; formal analysis, T.S.L.; data curation, T.S.L.; writing the original draft, T.S.L.; reviewing and editing, T.S.L. and K.R.R. All authors have read and agreed to the published version of the manuscript.

Funding: This research received no external funding.

Institutional Review Board Statement: The study was conducted in accordance with the Declaration of Helsinki and determined to be exempt from formal review by the Office for Research Protections at The Pennsylvania State University (Study 00012283).

Informed Consent Statement: Informed consent was obtained from all subjects involved in the study.

Data Availability Statement: Data sharing is not applicable to this article.

Conflicts of Interest: The authors declare no conflict of interest.

References

1. Associated Builders and Contractors, Inc. Construction Industry Faces Workforce Shortage of 650,000 in 2022. 2023. Available online: https://www.abc.org/News-Media/News-Releases/entryid/19255/abc-construction-industry-faces-workforce-shortage-of-650-000-in-2022 (accessed on 14 June 2023).
2. The Center for Construction Research and Training. CPWR Data Bulletin: Construction Employment Trends and Projections. March, 2022. Available online: https://www.cpwr.com/wp-content/uploads/DataBulletin-March2022.pdf (accessed on 14 June 2023).
3. Moskowitz, A.F.; Ramirez, M.R.; Schofield, K.E.; Ryan, A.D.; Zaidman, B.A.; McGovern, P.M. What are the risks to minors who work in the construction industry? *J. Occup. Environ. Med.* **2021**, *63*, e462–e463. [CrossRef]
4. Brown, S.; Brooks, R.D.; Dong, X.S. *Nonfatal Injury Trends in the Construction Industry*; CPWR: The Center for Construction Research and Training: Silver Spring, MD, USA, 2020. Available online: https://www.cpwr.com/wp-content/uploads/DataBulletin-December2020.pdf (accessed on 14 June 2023).
5. Love, T.S.; Roy, K.R. *Safer Engineering and CTE Instruction: A National STEM Education Imperative. What the Data Tells Us*; International Technology and Engineering Educators Association: Reston, VA, USA, 2022. Available online: https://www.iteea.org/SafetyReport.aspx (accessed on 14 June 2023).
6. International Technology and Engineering Educators Association. *Standards for Technological and Engineering Literacy: The Role of Technology and Engineering in STEM Education*; International Technology and Engineering Educators Association: Reston, VA, USA, 2020. Available online: https://www.iteea.org/stel.aspx (accessed on 14 June 2023).
7. Love, T.S.; Roy, K.R. Considerations for STEL-Aligned Professional Development Guidelines. In *Standards-Based Technology and Engineering Education*; Bartholomew, S.R., Hoepfl, M.C., Williams, P.J., Eds.; Springer Nature: Singapore, 2023. Available online: https://link.springer.com/book/9789819957033 (accessed on 3 July 2023).
8. Imperatore, C.; Hyslop, A. *2018 ACTE Quality CTE Program of Study Framework*; Association for Career and Technical Education: Alexandria, VA, USA, 2018. Available online: https://www.acteonline.org/professional-development/high-quality-cte-tools/high-quality-cte-development/ (accessed on 14 June 2023).

9.   Threeton, M.D.; Evanoski, D.C. Occupational safety and health practices: An alarming call to action. *Career Tech. Educ. Res.* **2014**, *39*, 119–136. [CrossRef]

10. Sandri, O.; Holdsworth, S.; Maslen, S.; Hayes, J. Contextualising capabilities for public safety in undergraduate sustainability engineering education. *Environ. Educ. Res.* **2023**, *29*, 451–472. [CrossRef]

11. Love, T.S.; Salgado, C.A. Teaching construction: A design-based course model. *Technol. Eng. Teach.* **2016**, *75*, 22–28.

12. Blankenbaker, E.K. *Construction and Building Technology*; Goodheart-Wilcox: Tinley Park, IL, USA, 2013.

13. Project Lead the Way. Engineering Curriculum. 2023. Available online: https://www.pltw.org/curriculum/pltw-engineering (accessed on 14 June 2023).

14. Hughes, A.J.; Merrill, C. Analyzing concrete beam design: Verifying predictions in T&EE classrooms. *Technol. Eng. Teach.* **2020**, *79*, 14–19.

15. Hughes, A.J.; Merrill, C. Concrete beam design: Pouring the foundation to engineering in T&E classrooms. *Technol. Eng. Teach.* **2020**, *79*, 8–13.

16. Hughes, A.J.; Merrill, C. Method of joints: Theory and practice of designing, building, and testing trusses. *Technol. Eng. Teach.* **2020**, *80*, 28–35.

17. Hughes, A.J.; Merrill, C. Solving concurrent and nonconcurrent coplanar force systems: Balancing theory and practice in the T&EE classroom. *Technol. Eng. Teach.* **2020**, *80*, 16–22.

18. Advance CTE. Architecture & Construction. 2023. Available online: https://careertech.org/architecture-construction (accessed on 14 June 2023).

19. National Association of Home Builders. Students Get Real-World Construction Experience in Their Own Parking Lot. 23 February 2023. Available online: https://www.nahb.org/blog/2023/02/students-get-real-world-construction-experience-in-parking-lot (accessed on 14 June 2023).

20. Maryland State Department of Education. Manufacturing, engineering, and technology. 2023. Available online: https://www.marylandpublicschools.org/programs/Pages/CTE-Programs-of-Study/Clusters/MET.aspx (accessed on 14 June 2023).

21. Love, T.S.; Roy, K.R.; Gill, M.; Harrell, M. Examining the influence that safety training format has on educators' perceptions of safer practices in makerspaces and integrated STEM labs. *J. Saf. Res.* **2022**, *82*, 112–123. [CrossRef]

22. Love, T.S.; Maiseroulle, T. Are technology and engineering educator programs really declining? Reexamining the status and characteristics of programs in the United States. *J. Technol. Educ.* **2021**, *33*, 4–20. [CrossRef]

23. Reed, P.A.; Ferguson, M.K. Safety training for career and content switchers. *Technol. Eng. Teach.* **2021**, *80*, 16–19.

24. Volk, K. The demise of traditional technology and engineering education teacher preparation programs and a new direction for the profession. *J. Technol. Educ.* **2019**, *31*, 2–18. [CrossRef]

25. Association of Career and Technical Education. Career and Technical Education Community Uniting to Address Teacher Shortages. Pearson, 21 March 2023. Available online: https://www.pearson.com/ped-blogs/blogs/2023/03/career-and-technical-education-community-uniting-to-address-teacher-shortages.html (accessed on 14 June 2023).

26. United States Bureau of Labor Statistics. Occupational Employment and Wage Statistics. 2023. Available online: https://www.bls.gov/oes/ (accessed on 14 June 2023).

27. Williams, T.O., Jr.; Ernst, J.V. Technology and engineering education teacher characteristics: Analysis of a decade of institute of education sciences nationally representative data. *J. STEM Educ. Innov. Res.* **2022**, *23*, 16–21.

28. Bowen, B.; Williams, T.O.; Napoleon, L.; Marx, A. Teacher preparedness: A comparison of alternatively and traditionally certified technology and engineering education teachers. *J. Technol. Educ.* **2019**, *30*, 75–89. [CrossRef]

29. Litowitz, L.S. A Curricular analysis of undergraduate technology & engineering teacher preparation programs in the United States. *J. Technol. Educ.* **2014**, *25*, 73–84.

30. Duncan, D.W.; Cannon, J.; Kitchel, A. Teaching efficacy: A comparison of traditionally and alternatively certified CTE teachers in Idaho. *Career Tech. Educ. Res.* **2013**, *38*, 57–67. [CrossRef]

31. Love, T.S.; Threeton, M.D.; Roy, K.R. A safety study on educators of technological and engineering design-based instruction in K-12 STEM related courses. *J. Technol. Educ.* **2023**, *35*, *in press*.

32. Love, T.S.; Roy, K.; Sirinides, P. A national study examining safety factors and training associated with STEM education and CTE laboratory accidents in the United States. *Saf. Sci.* **2023**, *160*, 106058. [CrossRef]

33. National Center for Construction Education and Research. OSHA-10 Training. 2023. Available online: https://www.nccer.org/learners-craft-pros/osha-10-training/ (accessed on 14 June 2023).

34. Cannon, J.; Tenuto, P.; Kitchel, A. Idaho secondary principals perceptions of CTE teachers' professional development needs. *Career Tech. Educ. Res.* **2013**, *38*, 257–272. [CrossRef]

35. Rose, M.A.; Shumway, S.; Carter, V.; Brown, J. Identifying characteristics of technology and engineering teachers striving for excellence using a modified Delphi. *J. Technol. Educ.* **2015**, *26*, 2–21. [CrossRef]

36. Lupton, G.T. Self-Identified Professional Development Needs of Virginia Career and Technical Education Teachers. Ph.D. Thesis, Virginia Tech, Blacksburg, VA, USA, 2021. Available online: https://vtechworks.lib.vt.edu/handle/10919/102411 (accessed on 14 June 2023).

37. Cannon, J.G.; Kitchel, A.; Duncan, D.W.; Arnett, S.E. Professional development needs of Idaho technology teachers: Teaching and learning. *J. Career Tech. Educ.* **2011**, *26*, 32–47. [CrossRef]

38. Bush, D.B.; Andrews, K. *Integrating Occupational Safety and Health Training into Career and Technical Education in Construction: Formative Research Findings*; CPWR: The Center for Construction Research and Training: Silver Spring, MD, USA, 2013. Available online: https://www.cpwr.com/wp-content/uploads/publications/Integrating_SandH_in_CTE_Training_Bush_2013.pdf (accessed on 14 June 2023).

39. Grytnes, R.; Grill, M.; Pousette, A.; Törner, M.; Nielsen, K.J. Apprentice or student? The structures of construction industry vocational education and training in Denmark and Sweden and their possible consequences for safety learning. *Vocat. Learn.* **2017**, *11*, 65–87. [CrossRef]

40. Chatigny, C. Occupational health and safety in initial vocational training: Reflection on the issues of prescription and integration in teaching and learning activities. *Saf. Sci.* **2022**, *147*, 105580. [CrossRef]

41. Nykänen, M.; Sund, R.; Vuori, J. Enhancing safety competencies of young adults: A randomized field trial (RCT). *J. Saf. Res.* **2018**, *67*, 45–56. [CrossRef]

42. Jones, J.; Belcher, G.; Elliott, K. Identifying technical competencies for architecture and construction education using the Delphi method. *Career Tech. Educ. Res.* **2021**, *46*, 3–15. [CrossRef]

43. Nichols, G.V. Analysis of the Safety Management Perceptions and Practices of Secondary Residential Construction Teachers in Vocational Education. Ph.D. Thesis, University of Arkansas, Fayetteville, AR, USA, 2000.

44. Love, T.S. Accident Occurrences and Safety Issues Reported by Mid-Atlantic P-12 Engineering Education Programs. In Proceedings of the Annual Conference and Exposition of the American Society for Engineering Education: Middle Atlantic Section, Harrisburg, PA, USA, 11–12 November 2022.

45. Sheskin, D.J. *Handbook of Parametric and Nonparametric Statistical Procedures*, 5th ed.; Chapman and Hall: New York, NY, USA, 2011.

46. Love, T.S.; Roy, K.R. Safety Issues and Accidents Associated with P-12 Pre-Engineering and Engineering Design Courses: Results from a National study. In Proceedings of the Annual Conference and Exposition of the American Society for Engineering Education, Baltimore, MD, USA, 25–28 June 2023.

47. Rigdon, E.E.; Feguson, C.E., Jr. The performance of the polychoric correlation coefficient and selected fitting functions in confirmatory factor analysis with ordinal data. *J. Mark. Res.* **1991**, *28*, 491–497. [CrossRef]

48. National Center for Education Statistics. Public High School Teachers of Career and Technical Education in 2007–2008. 2008. Available online: https://nces.ed.gov/pubs2011/2011235.pdf (accessed on 14 June 2023).

49. International Technology Educators Association. *Standards for Technological Literacy: Content for the Study of Technology*; International Technology Educators Association: Reston, VA, USA, 2000. Available online: https://www.iteea.org/67767.aspx (accessed on 14 June 2023).

50. Snyder, J.F.; Hales, J.A. *Jackson's Mill Industrial Arts Curriculum Theory*; West Virginia Department of Education: Charleston, WV, USA, 1981.

51. Occupational Safety and Health Administration. *Training Requirements in OSHA Standards*; Occupational Safety and Health Administration: Washington, DC, USA, 2015.

52. Roy, K.R.; Love, T.S. *Safer Makerspaces, Fab Labs, and Stem Labs: A Collaborative Guide!* National Safety Consultants, LLC.: Vernon, CT, USA, 2017.

53. North Carolina Department of Public Instruction. *School Science Facilities Planner*; Division of School Support: School Planning: Raleigh, NC, USA, 2013. Available online: https://files.nc.gov/dpi/documents/schoolplanning/science-facilities-planner.pdf (accessed on 14 June 2023).

54. West, S.S. Overcrowding in K-12 STEM classrooms and labs. *Technol. Eng. Teach.* **2016**, *76*, 38–39.

55. Stephenson, A.L.; West, S.S.; Westerlund, J.F.; Nelson, N.C. An analysis of incident/accident reports from the Texas secondary school science safety survey, 2001. *Sch. Sci. Math.* **2003**, *103*, 293–303. [CrossRef]

56. West, S.S.; Kennedy, L. Science Safety in Secondary Texas Schools: A Longitudinal Study. In Proceedings of the Twelfth Annual Hawaii International Conference on Education, Honolulu, HI, USA, 5–8 January 2014.

57. Maximum Class Size, 8VAC20-120-150, Virginia General Assembly. 2012. Available online: https://law.lis.virginia.gov/admincode/title8/agency20/chapter120/section150/ (accessed on 14 June 2023).

58. National Fire Protection Association. *NFPA 101, Life Safety Code: 2021 Edition*; National Fire Protection Association: Quincy, MA, USA, 2020.

59. Threeton, M.D.; Kwon, K.; Fleck, J.A.; Ketchem, R.B.; Farzam, L. An investigation of instructional practices which promote occupational safety and health. *Int. J. Occup. Saf. Ergon.* **2019**, *27*, 902–910. [CrossRef]

60. Occupational Safety and Health Administration. Hand and Power Tools. 2023. Available online: https://www.osha.gov/hand-power-tools (accessed on 14 June 2023).

61. International Technology and Engineering Educators Association. Safety in Technology and Engineering Education; 2023. Available online: https://www.iteea.org/Resources1507/Safety.aspx (accessed on 14 June 2023).

62. Power Tool Institute, Inc. Power Tool Institute. 2023. Available online: https://www.powertoolinstitute.com/ (accessed on 14 June 2023).
63. Virginia Tech Department of Agricultural, Leadership, and Community Education. Laboratory Safety Resources. 2023. Available online: https://www.alce.vt.edu/signature-programs/team-ag-ed/teacher-resources.html (accessed on 14 June 2023).

sustainability

Article

Bringing Project-Based Learning into Renewable and Sustainable Energy Education: A Case Study on the Development of the Electric Vehicle *EOLO*

Jonathan Álvarez Ariza [1,*] and Tope Gloria Olatunde-Aiyedun [2]

[1] Department of Technology in Electronics, Engineering Faculty, Corporación Universitaria Minuto de Dios-UNIMINUTO, Bogotá 111021, Colombia
[2] Department of Science and Environmental Education, University of Abuja, Abuja 900105, Nigeria; tope.aiyedun@uniabuja.edu.ng
* Correspondence: jalvarez@uniminuto.edu

Abstract: In recent years, there has been a growing interest in education for sustainable development (ESD). Although several national and international agencies, e.g., the UN or UNESCO, have promoted its deployment in higher education institutions, educators are still facing problems with how to articulate this type of education within the curriculum, allowing students to develop their technical and labor competencies, and soft skills as well. In this way, this study describes a methodology with Project-Based Learning in renewable and sustainable energies through the development of an electrical vehicle (EV) known as *EOLO*. This initiative arose from an industry-academia collaboration to develop the first Colombian EV with the support of solar and wind energy sources. Twelve engineering students participated in the development of the vehicle through a set of capstone projects over a year and a half with the support of two tutors (professors) and two engineers (technical staff) of the project. Additionally, two versions of *EOLO* with vertical and horizontal axis wind turbines were made with the cooperation of the students. The results evidence that the methodology helped to engage students, promoted meaningfully and situated learning through real-world problems in renewable energies, and fostered motivation and peer collaboration. Nonetheless, aspects such as the improvement of the communications channels, the revision of the complexity of the projects, the sense of community to achieve a common goal, or the tutoring and monitoring processes should be strengthened for further initiatives and/or active learning methodologies. In this sense, some challenges and recommendations that can help to develop methodologies that combine ESD and engineering are provided based on the experience in this study.

Keywords: project-based learning; education for sustainable development; renewable energy; engineering education; electric vehicle; industry-academia collaboration

Citation: Ariza, J.Á.; Olatunde-Aiyedun, T.G. Bringing Project-Based Learning into Renewable and Sustainable Energy Education: A Case Study on the Development of the Electric Vehicle *EOLO*. *Sustainability* **2023**, *15*, 10275. https://doi.org/10.3390/su151310275

Academic Editors: Clara Viegas and Natércia Lima

Received: 19 May 2023
Revised: 22 June 2023
Accepted: 23 June 2023
Published: 29 June 2023

1. Introduction

During the last decade, the interest in Education for Sustainable Development (ESD) has increased considerably in the engineering curricula and it has inspired the reformulation of the role that engineers have in society and the impact that their products, designs, and services could have on the environment. Additionally, the Sustainable Development Goals (SDG) promoted by UNESCO [1–4], especially, SDG 4 (*quality Education*), SDG 7 (*affordable and Clean Energy*), SDG11 (*sustainable cities and communities*), and SDG13 (*climate action*) have accelerated this trend. Then, aspects such as the ability to determine the most appropriate renewable energy strategy for a community, the understanding of the importance of lifelong learning, the reduction of educational gaps and inequalities, or the capability to develop a reliable plan for sustainable energy production are announced as examples of key factors to achieving the SDGs [1,3].

Similarly, the need for responsible and eco-friendly engineering is continuously included and revisited in the assessment criteria of the accreditation agencies such as the Accreditation Board for Engineering and Technology (ABET) and the Engineering Council (UK-SPEC), which are the base to formulate engineering programs at the levels of undergraduate, master, and Ph.D. degrees in the US, the UK and in other higher education institutions (HEIs) around the world. For instance, the student outcomes in criterion 3 of the ABET (2020–2021) [5] indicate that graduate engineering students should have "an ability to recognize ethical and professional responsibilities in engineering situations and make informed judgments, which must consider the impact of engineering solutions in global, economic, environmental, and societal contexts", as well as "an ability to apply engineering design to produce solutions that meet specified needs with consideration of public health, safety, and welfare, as well as global, cultural, social, environmental, and economic factors". By the same token, the European Commission [6] states that the role of HEIs is to educate critical thinkers, problem solvers, and agents of change to be responsible citizens. For that, HEIs need to engage students in complex analysis of reality and create spaces to foster democratic participation and decision-making processes. Although around 1400 HEIs have endorsed, signed, and approved plenty of declarations about Sustainable Development (SD) from 1990 until now mainly in Europe and US [7,8], questions still remain about how engineering curricula can contribute to SD and how these curricula could keep students motivated and engaged. In particular, these issues have been reported in numerous studies in engineering education [9–14]. The authors claim in these studies for a need to understand engineering holistically and not only from the traditional technical point of view regarding SD. Also, as indicated in the study [9], in which SD was analyzed through the engineering curricula of 25 Spanish universities, the biggest identified challenge was the ensuring of SD competencies as a transversal initiative and not only from the isolated efforts of the academic staff to embed SD into the engineering curricula.

Nonetheless, as a counterpart, we also think that the contributions of Educators and Engineers from the classrooms are important to ensure effective inquiry and reflection about SD with the construction of methodologies, courses, and frameworks, among others, in this regard. With these contributions, the engineering curricula and policies in HEIs could be gradually changed and transformed to benefit the real integration and understanding of SD from the standpoint of students, teachers, and stakeholders in the education field. Then, the main purpose and motivation of this study were to create an educational methodology that responded to the learning needs of the students regarding sustainable energy education (solar and wind), and embedded and power systems in the design of the electric vehicle *EOLO*. This was a project in an industry–academia collaboration to design and deploy an electrical vehicle entirely produced in Colombia to contribute to the e-mobility in our country. In particular, this study was focused on the SDGs (4: Quality education, 7: Affordable and Clean Energy, and 9: Industry, Innovation, and Infrastructure). For the development of *EOLO*, ($n = 12$) students were engaged in the methodology of Project-Based Learning (PjBL or PBL) as part of a set of capstone projects. In addition, the 12 students received the accompaniment of two tutors and a continuous follow-up from the technical staff of the project *EOLO* for one year and a half. During this period, two versions of *EOLO* were constructed with the work, efforts, and cooperation of the students. The students were divided into small groups of two or three people with specific tasks and purposes to attain the proposed goals in *EOLO* and foster peer collaboration. The different problems addressed through the PjBL methodology allowed the students to inquire, design, develop, and deploy solutions, employing the skills and competencies acquired along their learning process. Although it seems that the number of students is small, this was because the amount of projects in *EOLO* was limited and one feature of the PjBL methodology is the work usually in small groups to achieve the proposed goals (educational and technical).

To evaluate the methodology and the experience, a survey with 19 questions was provided to the students. Additionally, a second survey was produced with 13 questions for the two technical leaders of *EOLO* in order to complement the standpoint of the students and identify recommendations and aspects to improve in the methodology. With the previous approach, we search to address the lack of methodologies that help to embed the SDGs in the curricula of engineering, especially those mentioned above. Also, we want to describe in-depth the details and results of the methodology, which can benefit other educators in the construction and/or incorporation of active learning methodologies in the areas of renewable energy, sustainability, embedded systems, and electrical vehicles, among others. It is worth mentioning that the results presented here expand on those presented in the conference paper [15].

This paper is divided into the following sections. Section 2 shows the related works. Section 3 describes the details of the *EOLO* project, participants, instruments, and the methodology including some technical factors. Section 4 outlines the results and discussion of the methodology from the standpoint of the students, tutors, and main staff of the *EOLO* project, including some representative technical results. Section 5 exposes several educational challenges and recommendations that were identified in the process. Finally, Section 6 depicts the limitations, further work, and conclusions of this study.

2. Related Works

2.1. Educational Works

Several educational methodologies have been described to foster ESD, mainly in HEIs. For instance, Cebrián et al. [12] expose four student-centered methodologies, namely, project or problem based-learning, case studies, simulations, and cooperative inquiries. The authors employed both a literature review and an expert analysis to determine the importance of each one of the previous methodologies in the process of ESD. Authors conclude that the *smart classrooms*, that is, the learning settings that use the methodologies indicated above can foster collaboration, knowledge sharing, and successful interaction between students and teachers. Similarly, the authors in [16] proposed a project-based learning methodology for non-technical students with different laboratory experiments in the fields of control and actuator technologies and e-mobility. The authors suggested that the methodology can help to *bring to life* the discourse about ESD and it allowed its incorporation inside interdisciplinary projects in electrical engineering. Also, a methodology for the development of eco-design and eco-efficiency competencies in engineering is presented in [13]. In the methodology participated 49 students during 2018–2019 who performed three assignment stages: Identifying the product and fabrication processes, eco-design analysis of the product and eco-efficiency evaluation, and rethinking the product to decrease its environmental impact and promote circularity. In the conclusion is highlighted the importance of the interaction between students and industrial agents to create new ways for collaborative projects. In the same way, Curiel-Ramirez et al. [17] propose a new curriculum for automotive engineering education. Through the analysis of the topics regarding advanced driver-assistance systems (ADAS), product design, and chassis, among others, in several universities around the world, the researchers identified the key factors for the new curricula.

Additionally, an approach focused on inquiry-based learning for renewable energy sources (RES) is described in [18]. The approach helped students with low academic performance and it engages them in the topics of RES through problem-solving activities that address some issues in the current renewable energy panorama. In addition to this approach, the authors in the work [14] discuss an eight-year experience teaching aerodynamics, mechanics, and hydraulics, which were aligned with the SDGs, critical thinking, and problem-based learning. The authors depict the different phases such as definition and planning, monitoring and execution, assessment, and group and learning activities. The results of the methodology evidenced that the activities generated reflective thinking and a cooperative learning environment about four SDGs (SDG4: Quality education, SDG13: Climate change, SDG7: Affordable and clean energy, SDG 6: Clean water and sanitation). Alternatively, an educational strategy for ESD through a kit for energy, environment, and sustainability in Guinea-Bissau is proposed in [19]. The authors indicate the experience to use the kit called KEAS with several teacher trainers in the Guinea-Bissau School of Education. The authors mention several principles for the design of the kit such as being culturally appropriate, user-friendly for students and teachers, low-cost, or sensitive to gender issues. In addition, with the kit, teachers can explore practical work in science classes, helping to enhance the learning process of the students.

Although the previous methodologies are successful cases of the incorporation of ESD principles in an international perspective, in the Colombian context, however, there are few examples of methodologies in engineering education that help to understand the implications of the ESD both into the curricula and the educational pathway of the students. Thus, some examples of national policies and related works have been exposed in the studies [20–23], but still in some of them it is not clear how the SDGs can be integrated properly into the technical competences or abilities that engineering students should develop. This may be due in part to the fact that the SDGs are viewed in a general way in courses that are not an active part of the engineering curricula. For example, in these courses, students can address environmental care through recycling activities, water care, sanitation, fundamentals of SGDs, etc., but one issue here is how to properly integrate these concepts into the main corpus of technical abilities or competencies that one student must have in engineering and more specifically in this work, in electronic or electrical engineering or technology. Some strategies have been elicited to overcome this issue in higher education (HE), which include training through lectures, involving students, individual interaction, project-based and inquiry-based learning, and promoting personal and collaborative learning along the sustainability context [12,24,25]. In addition, HEIs should be leaders in academic changes that include the ability to reformulate curricula that respond to the SDGs through initiatives such as linking ESD pedagogies to special literacies, e.g., in science, reading, or mathematics, digesting new sustainable thinking and practice in the industry/profession, and rethinking quality learning outcomes under the lens of ESD [26].

Complementarily, this study was shaped by a work in progress (WIP) presented to the IEEE Frontiers in Education Conference (FIE) [15]. In this preliminary study, we exposed the results of the students' perceptions regarding the methodology forwarded in the first version of *EOLO*. At this point, it is important to mention that two versions of the EV—*EOLO* V_1 and *EOLO Pro*—were built with the support of the students during the years 2017–2020. The conclusions of this first work show that students were engaged in the methodology and learned concepts about renewable energy sources, CAN-Bus communications, electrical wiring for EVs, and embedded programming. For this reason, e.g., the 75% of the students stated that the projects increased their motivation for learning concepts and applying them in the deployment of their solutions in the EV. Similarly, at this point, the students manifested the need for reinforcing the tutoring process with the teachers and technical staff of *EOLO*. These suggestions were taken into account in the construction of the second version of *EOLO* called *EOLO Pro*.

2.2. Technical Works

At the technical level, some related works were deemed, even though the interest in the current study is not technical but instead educational. For instance, in the study [27], the author describes a hybrid vehicle that combines a photovoltaic (PV) module with a micro wind turbine. During sunny days, the PV module and the turbine can add 19.2 km of extra autonomy to the vehicle. Nonetheless, this autonomy is achieved at the cruise speed of 121 km/h. A second experimental work is presented in [28]. In this work, the author simulates the behavior of a car with a vertical axis turbine, employing computational fluid dynamics (Autodesk CFD). The author emphasizes the shape of the turbine blades as a factor that influences the pressure distribution on the car and can cause fatigue in the vehicle's parts in the front and rear. Another experimental study about the possibility to install a wind turbine in an EV is depicted in [29]. In the simulations, the wind turbine with a voltage generator can provide an electrical power of 3.26 kW when the car is moving at a speed of 120 km/h. Finally, in this work, the authors recommend further work to identify how the power generated could be increased. In reference [30] is shown a rapid prototyping technique with 3D printing to create a wind turbine focused on vehicles and domestic power generation. The study exposes the feasibility to implement and deploy a vertical wind turbine with PLA with biodegradable properties and a duration of 20 years. Moreover, the simulations show that the turbine can be used in vehicles up to the speed of 80 km/h. Previous works are meaningful examples of the feasibility to create and install a wind turbine in a vehicle for electrical power generation. Nevertheless, there are some points to improve in these works. The first one is the need for the vehicle to reach the cruise speed of 120 km/h as claimed in the studies [27,29], which could not be optimal for urban traffic. The second consideration is that many of these works employ simulations as proof-of-concept and do not show real implementations in EVs. Searching to address this technical gap is just how the project *EOLO* arose. The vehicle's concept was focused on the construction of a wind turbine that was feasible for EVs, and on the other hand, on features of the EV to gain autonomy with the drop-off of the wind resistance thereof, e.g., creating different wind ducts in the front part of the EV. The tests of the vertical wind turbine installed in the vehicle, which were performed directly in a racetrack of 667 m long, completing 110 laps, can be consulted in the study [31]. The main results in the tests indicate that the wind turbine can provide electrical power over the vehicle's speed of 50 km/h, and allows increasing the vehicle's autonomy between 4% to 8% considering that the maximum boundary of power generated was 1289.59 W and the vehicle consumption was 14,529.9 W. For researchers interested in the performed technical simulations and tests (see work [31]).

3. Context and Method

3.1. EOLO Project

The *EOLO* project arose in 2015 as an initiative to create an electrical vehicle 100% designed and produced in Colombia. The project was released by the Corporación Industrial Minuto de Dios (CIMD), a Colombian entity that aims to serve the entrepreneurial and industrial development and income generation, through the protection and generation of employment, the dynamization of entrepreneurship and training for productive qualification [32]. *EOLO* was the first electric vehicle (EV) developed in Colombia with a proprietary wind charging mechanism through Savonious turbines [33]. These devices can be installed in small places, e.g., in the front part of the vehicle, and allow us to propel voltage generators to charge a set of installed lithium-ion batteries. The outcome of the incorporation of these devices is an increase in the vehicle's autonomy between 4–8% in the *EOLOs* case [31]. Two *EOLO* versions known as *EOLO* V_1 and *EOLO Pro* were designed and constructed between 2017–2020, and their appearances are depicted in Figure 1. The first version of *EOLO* has a vertical axis wind turbine (VAWT), meanwhile the second one has a horizontal axis wind turbine (HAWT) to save space and weight in the EV. Additional information about the vehicle can be found in the references [31,34–37]. The main technical details of both versions can be seen in Table 1.

As described, the Savonious turbines were installed vertically as Figure 1a,b depict for *EOLO* V_1. Concerning *EOLO Pro*, the turbines were redesigned for searching less weight and better improvement of the autonomy of the vehicle, which is still under development. In this way, turbines were installed horizontally. In addition, for this last version, a second photovoltaic system with a nominal capacity of 118 W was installed on the vehicle's roof and in its rear part. Six panels were installed with a total capacity of 100 W (12 V–7.5 A) for three of these in the roof, and 18 W (12 V–1.5 A) for three of these in the rear part of the EV as Figure 1d depicts. The goal of both systems (solar and wind) was to improve the vehicle's autonomy by a maximum of approximately 11% (8–10% wind + 0.6–0.81% solar) considering the power vehicle's consumption mentioned above. Nonetheless, this improvement in autonomy is currently under revision and development. In this way, for the design, monitoring, and deployment of the energy and the electric systems for both versions of

EOLO that included, e.g., lighting, communication, and control systems, the CIMD made an industry-academia collaboration with the university Corporación Universitaria Minuto de Dios-UNIMINUTO in 2018. Following this alliance, two professors and twelve students from the engineering faculty were engaged in the project. Thus, according to several meetings held between the group of professors and the technical staff of the *EOLO* project to gather technical requirements, six projects were generated, which will be discussed in the next sections.

Table 1. Technical parameters of both *EOLO's* versions. Notice that in both versions of *EOLO* the same electric motor was employed.

Parameter	EOLO V_1	EOLO Pro
1. *Motor reference and power (HP)*	HPEVS AC51-88 HP	HPEVS AC51-88 HP
2. *Torque (ft-lbs)*	108 ft-lbs	108 ft-lbs
3. *Power system voltage (V)*	144 V	144 V
4. *Battery Type*	12 V-LiFePO$_4$	12 V-LiFePO$_4$
5. *Number of batteries*	48	48
6. *Max. speed (km/h)*	120 km/h	120 km/h
7. *Charging time (hrs.)*	8 h (120 V)–5 h (220 V)	8 h (120 V)–5 h (220 V)
8. *Weight (Kg)*	1200 Kg	800 Kg

<table>
<tr><td align="center">(a)</td><td align="center">(b)</td></tr>
<tr><td align="center">(c)</td><td align="center">(d)</td></tr>
</table>

Figure 1. Appearance of the different *EOLO's* versions. Savonious turbines are located in the front part of the vehicle. (**a**) *EOLO* V_1. Front view; (**b**) *EOLO* V_1. View of the VAWT mechanism; (**c**) *EOLO Pro*. HAWT is located at the front; (**d**) *EOLO Pro*. Photovoltaic system on the vehicle's roof and rear.

3.2. Context and Participants

The students and professors who participated in the project belong to the program of Technology in Electronics at the Colombian university Corporación Universitaria Minuto de Dios-UNIMINUTO. In Colombia, technological careers are understood as types of programs focused on labor fields, with a duration between two and three years. Most of the subjects established in the curricula of those programs are focused on the necessary abilities that a determined industrial or productive sector requires, e.g., in some programs of technology in electronics, their emphases include telecommunications, industrial communications, automation, or bioengineering [38]. The students had a range of ages between 17–35 years old and they were between the fourth and sixth semesters of their careers in Technology in Electronics. Hence, the students already possessed knowledge of embedded systems, circuits, or the fundamentals of power electronics. Nonetheless, as we shall see, several additional technical concepts were developed or reinforced with the group of students in these areas. Furthermore, eight students worked during the day and took their classes in a night schedule (6:00 p.m.–10:00 p.m.) and typically performed the different tasks in the project on Saturdays. Besides, two engineers responsible for the *EOLO* project monitored the students throughout the process of the accomplishment of the proposed technical tasks. The students invested approximately between 8–12 h per week on the tasks assigned to each project.

3.3. Educational Objectives

Three educational objectives (EOs) were constructed for the different activities performed by the students. These are listed as follows:

- **EO1**: Reinforce and strengthen the learning about sustainable energy sources (wind and solar), embedded, and power systems with the design and deployment of technical solutions for the projects proposed in *EOLO*.
- **EO2**: Motivate and engage the students to create solutions eco-friendly and eco-responsible from the technical standpoint regarding e-mobility and sustainable energy sources.
- **EO3**: Promote teamwork and the development of soft skills in the students that help to fulfill the goals and aims of *EOLO*, which includes enhancing the educational process of the students.

Our primary objective was to provide high-quality education to our students with a representative project in e-mobility that allows them to put into practice the knowledge and learning developed in their educational pathway in the areas of sustainable energy sources, embedded, and power systems. In this sense, this aim is aligned directly with the (**SDG4**: Quality education), but the different actions of the students in *EOLO* also are aligned with the (**SDG7**: Affordable and clean energy), and (**SDG13**: Climate change). In addition, in the normal curriculum of the career of Technology in Electronics, the topics related to sustainable energy systems are not encompassed. Therefore, throughout the project, the students reviewed these topics and applied them with hands-on activities, which is a key point in engineering and technology education.

3.4. Description of The Capstone Projects

As described, six projects were outlined according to the technical requirements in *EOLO*. The projects combine embedded systems, handling of power electronics, CAN-Bus communications, Human Machine Interface (HMI), and battery balancing. The goal of the projects in *EOLO* was to employ the power generated by the wind turbine and the solar panels in several *secondary* systems like the previous ones without consuming the power of the *primary* system that provides power to the electric motor, whose nominal capability is 15 KW. Moreover, the remaining generated electric power can be used to charge the main battery system of the EV.

The projects were continuously revisited in complexity and relevance by the professors and main technical staff involved in the project before assigning them to the students. Table 2 shows the list of these projects with their technical description. The formulated projects responded to the following requirements:

- Reduction of the power consumption of the lighting system. The first version of this system had a consumption of 300 W with halogen lamps, which made it unfeasible.
- A balancing system for the batteries. Without this system, the batteries could be damaged. The cost of each battery is approximately USD 250 and the vehicle needs 48 of these.
- A HMI system since the first version of *EOLO* did not have it. The system provides different functions to the driver in order to control some features of the vehicle. If it is possible, the system should have a touch screen for some of these functions.
- A system with CAN-Bus protocol (typically used in automotive engineering) to communicate the different units or nodes in the vehicle and reduce electric wiring.
- Redesign the electric wiring according to the normative in vehicles [39] to optimize the number of conductors needed in the vehicle.
- Introduction of the Savonious turbine and the solar PV panels in the recharging system of *EOLO pro*.

Table 2. List of formulated projects for *EOLO*.

Project	Description
1. *Balancing batteries*	Balancing system for at least 4 Lithium Iron Phosphate batteries (LPF) ($LiFePO_4$). These batteries serve as the power supply for the electric engine of the EV.
2. *Electric windows and mirrors*	Current sensing and control system for the mirrors and the electric windows.
3. *Redesign of electric wiring*	Redesign of the electric wiring according to the technical requirements.
4. *Lighting system front and rear*	Lighting system with LED lamps controlled by CAN-Bus protocol.
5. *Electric mechanism for vehicle's doors*	Electric mechanism to open and close the vehicle's doors.
6. *Driver-focused Human Machine Interface (HMI)*	HMI with a touch screen that supports functions such as tachometer, battery status, command buttons for light intensity (high, medium, low, strobe) and stationary lights.

With the previous requirements, at the technical level, the projects were oriented towards the architecture presented in Figure 2 for the electric system of the vehicle. Here, by now, the function of turning on the reverse lights is not available. The systems presented in this figure are interfaced to a *second battery* independent of the primary set of batteries of the EV, which is charged with the wind turbine and the solar panels. In the schematic of Figure 2, the devices and components were selected initially by the professors and engineers of the *EOLO* project. The hardware devices such as dsPICs, which are Digital Signal Controllers (DSCs) [40] for the design of the different nodes in the vehicle, including the Electronic Control Unit (ECU), the touch screen with the FT900Q microcontroller for the HMI [41,42], or the low-cost pulse width modulation (PWM) controller FL7760 [43] that controls the intensity of the front LED lamps, were selected according to parameters such as energy consumption, price, robustness, scalability, easiness to program, and availability of educational and technical resources to learn and understand their functioning. Notice that the different nodes, e.g., the ECU and the nodes for the front and rear lights, are communicated through a CAN-Bus protocol. This type of communication is not normally viewed in the courses on embedded systems in our curricula, for this reason, the fundamentals of its functioning and its importance in automotive systems were taught to all students who participated in the methodology. In parallel, other additional concepts in renewable energy, photovoltaic systems, embedded programming of DSCs, and functioning of electrical vehicles were introduced to the students since these were employed transversely in all projects in Table 2. Additionally, these concepts were reinforced by the tutors in the project, and the students had the necessary educational materials.

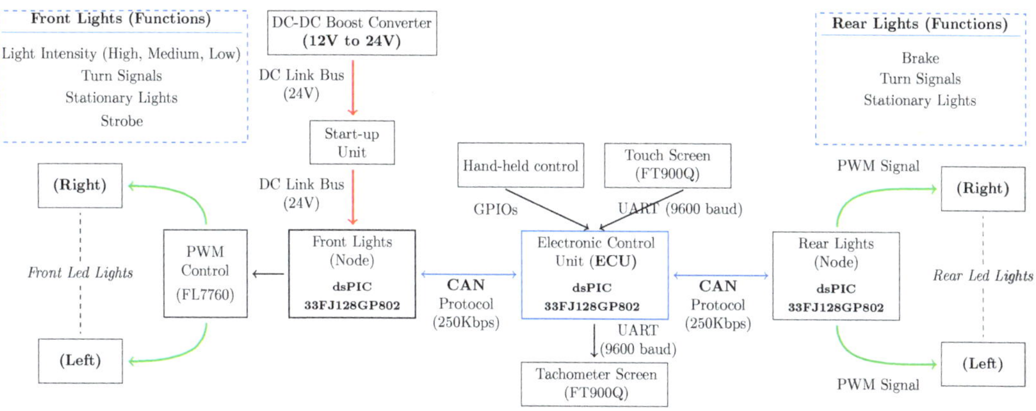

Figure 2. Proposed architecture of the electric system for both versions of *EOLO*. Kbps: Kilobytes per second, PWM: Pulse Width Modulation, GPIO: General Purpose I/O, UART: Universal Asynchronous Receiver-Transmitter. Red: DC Link Bus (12V, 24V), blue: CAN-Bus communication, green: PWM signals for the lighting system.

3.5. Educational Materials

As a technical requirement, the students must incorporate dsPIC devices in the creation of each node. We selected these devices due to some features such as processor speed, power consumption, and the number of protocols including CAN-Bus, or remapping pins. The programming of the dsPIC devices is not a current topic in our curricula. Then, the professors decided to create some utilities and educational materials based on prior works. For instance, for the programming of the dsPICs, the tutors created a utility through block-based programming called *DSCBlocks*. Through this application, the students can create codes using graphical blocks and can directly program the dsPIC with them to handle different peripherals such as (CAN-Bus, UART, PWM) or ports thereof. The students at any moment can see the respective code for the blocks in C language for the architecture of the dsPIC. Additionally, the supporting educational materials to understand the 16-bit dsPIC architecture were provided to the students before the stage of design. The overall appearances of the application and the educational materials are depicted in Figures 3 and 4, respectively. Detailed information about the application *DSCBlocks* can be found in the reference [44], and in the related study [45]. The e-learning tool Xerte [46] was employed as a tool for the construction of the educational materials that started with the concepts of the dsPIC architecture and finished with the features of the application *DSCBlocks*. Xerte provides a complete environment to develop materials with interactive resources such as videos, quizzes, games, or web pages that can accompany the learning process of the students. Other educational materials in solar photovoltaic panels, battery management, and electric vehicles were provided to the students timely with Xerte.

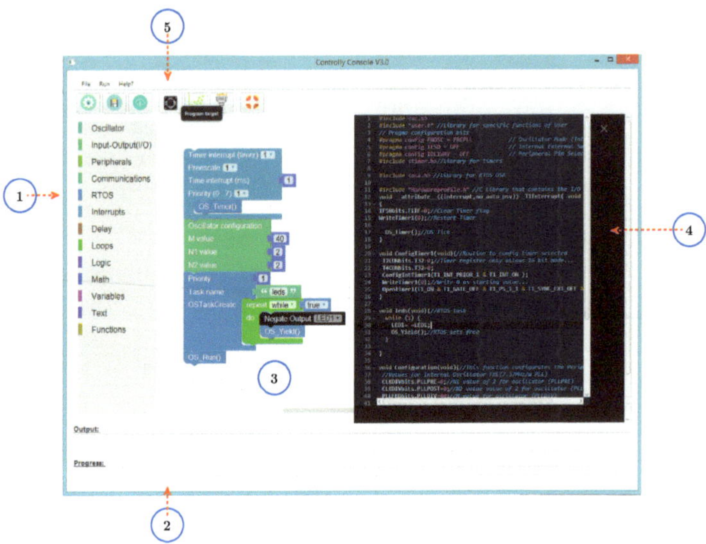

Figure 3. Overall components of the User Interface (UI) of *DSCBlocks*: (1) blocks palette; (2) console output; (3) working area; (4) real-time code tab; and (5) toolbar.

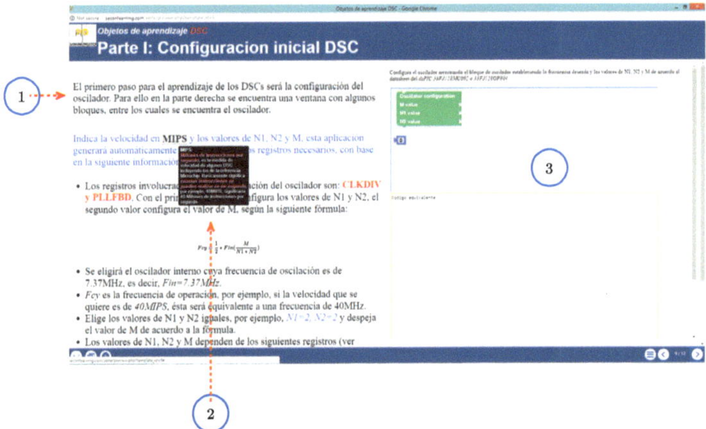

Figure 4. Educational materials in Xerte (Spanish version). (1) example of educational material; (2) glossary option of Xerte; and (3) embedded resource (interactive web page) with graphical blocks.

3.6. Methodology

The educational methodology was structured according to the guidelines of the active learning methodology *Project-Based Learning* (PBL or PjBL). PjBL is defined in terms of an interdisciplinary, student-centered, and constructivist approach that enables collaboration and reflection within real-world practices [47–49]. Among the advantages of PjBL are high-lighted learning of students or teachers, collaboration and sense of community, motivation, student-centered learning, and versatility for education [47]. According to Dumitrache et al. [50], the projects should be formulated considering aspects such as relevance for the discipline, complexity, attractiveness, and simplicity. Also, the students must know how they will be evaluated and the evaluation criteria. Then, several *capstone projects* were generated between the technical staff of *EOLO* and professors since their inclusion has been demonstrated to be a catalyst of meaningful learning, peer collaboration, and motivation in engineering because students should use the learning and knowledge developed along their education pathway [51]. Capstone projects and courses constitute the culminating experience in engineering curricula [52]. Thus, in this case, the methodology in these projects had the following overall stages:

1. Project formulation.
2. Project selection.
3. Project design, development and deployment.
4. Project debugging.
5. Presentation of final results.

Each one of these stages will be explained in the following subsections. It is important to indicate that the stages of project design, development, deployment, and debugging had a continuous follow-up and feedback by the professors and technical staff of *EOLO*. In this way, any design created by the students was implemented and deployed in both versions of *EOLO*. If an error or malfunction was detected, the feedback of the staff mentioned, allowed the students to return to the stage of design to check the problem and found the root of this to fix it, and again deploy the solution.

3.6.1. Stage 1: Project Formulation

The projects were formulated according to the technical requirements of *EOLO*. As described, multiple meetings were held to formulate and coordinate the projects between the professors and the engineers in *EOLO*. Since the requirements were defined previously by the technical leaders of *EOLO*, firstly, professors evaluated the pertinence, relevance, and complexity of these requirements for the students according to an examination of the required competencies in the function of the curricula. Secondly, with this examination, the professors determined other additional concepts that the students should address in the development of the technical tasks. Then, each project of Table 2 was formulated and conditioned according to the abilities that the students must display under the lens of the curricula. In parallel, as shown, 12 students that manifested their intention to contribute to the *EOLO* project were selected to participate in it.

3.6.2. Stage 2: Project Selection

The students were asked to choose the project more convenient under their abilities. Moreover, the students selected their peers in the project and worked in groups of two or three people. Only a group of one person was allowed since the tasks and complexity of this project were enough for a student. Also, the students had several meetings with the *EOLO* staff to know the technical requirements and expand the perspective of the projects.

3.6.3. Stage 3: Project Design, Development and Deployment

In this stage, the students started with the design of their solutions for each project. In time slots of 20 days, the students exposed their advancements to the professors and main engineers of *EOLO* in a short document and presentation. Overall, this stage had the steps described in Figure 5, which started with the activity of brainstorming and finished with the implementation and testing of the solutions proposed by the students.

In the *brainstorming* part, the students by group proposed several solutions for their projects. Afterwards, these solutions were reviewed and evaluated initially by the group of professors/tutors that accompanied the process. Then, based on the feedback received, the students chose the most suitable solution considering aspects such as cost, the feasibility of implementation, energy consumption, and the availability of hardware devices in local stores.

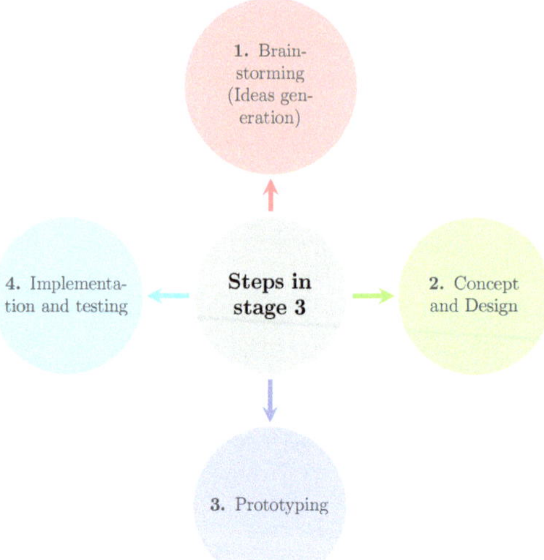

Figure 5. Steps for the stage of Project design, development and deployment.

With the solution proposed and the evaluation of its feasibility, the students started with the process of *concept comprehension and design*. Regarding concept comprehension, the students understood the additional concepts needed to make the designs of their solutions. In the design were deemed technical aspects such as power consumption, power saving modes for the DSCs, processing speed, data analysis of current and voltage for the solar PV panels and the wind generator, distance between the nodes to perform the CAN-Bus communication, the electric noise, the distribution of the HMI, among others.

Once the designs were concluded, the students started with the step of *prototyping*. In this, devices such as a 3D printer, Arduino, some development boards for DSCs, and the touch screens for the HMI were used to speed up this step and to test the prototypes quickly in the versions of *EOLO*. Some examples of tasks made by the students were: Programming basic commands (On-Off) in dsPIC, reading and writing data in the Universal Asynchronous Receiver-Transmitter (UART) peripheral, electric modeling of mirrors and windows, temperature sensing in Arduino, modeling consumption of led lamps, configuration of basic buttons in screens, PWM control, etc. In the second part of the prototyping stage, the students improved their designs and they built the respective printed circuit boards (PCBs) and their housings in a 3D printer in the function of the technical requirements.

The students *tested* their prototypes directly in the versions of *EOLO*. Nonetheless, the *EOLO* V_1 was transformed in a laboratory as the students performed the different tests, and these prototypes were quickly adapted to the *EOLO Pro* version. The advancement in the prototyping and its implementation was continuously monitored by the tutors and engineers of *EOLO*. Over 1250 h of tests were employed in the different nodes that were developed in this stage in continuous work with the students.

3.6.4. Stage 4: Project Debugging

Any error presented in the design and implementation was analyzed jointly by students, tutors, and engineers to find the source of the failure and provide a timely solution to the problem. Then, students made the corrections to their designs, executing the steps in Figure 5 again.

3.6.5. Stage 5: Presentation of Final Results

When the student finished their designs and their implementation in both versions of *EOLO*, they exposed their results in a report and presentation before the community and some academic peers. With this task, the students completed their training within the *EOLO* project. Nonetheless, as indicated, students submitted every 20 days a brief report with their activities and advances. Some figures that illustrate the previous stages are depicted in Figure 6.

Figure 6. Procedure followed in the methodology with the students. (**a**) Meetings with students; (**b**) technical requirements gathering; (**c**) explanation of concepts; (**d**) first designs (lighting system); (**e**) first designs Photovoltaic (PV) system; (**f**) implementation and testing; (**g**) implementation and testing (HMI system); (**h**) some students and members of the *EOLO* project; and (**i**) implementation and testing of lighting system in *EOLO Pro*.

3.7. Instruments

In order to assess the methodology, a survey with 19 questions (13 closed-ended, 6 open-ended) was applied to the students. The survey sought to know the perception of the students concerning learning, motivation, problems, and recommendations with the methodology. The students' survey was expanded in the present study regarding its first version whose results are consigned in the prior study [15]. Also, for this new instrument, the closed-ended questions were complemented and divided into three sections, the first one with 8 questions (Q1–Q8) and Cronbach's alpha (α) = 0.88, ranged from (1) Strongly disagree to (5) Strongly agree. The second one with 3 questions (Q9–Q11), ranged from (1) Very poor to (5) Excellent. One question (Q12) assessed the difficulty of each project in the range of (1) Very easy to (5) Very difficult. In the last one (Q13), the students provided a global evaluation of the methodology in the range of 0–5. The value of (α) was increased for the first version of the survey consigned in [15] from (α) = 0.71 to (α) = 0.88 for the current survey, which demonstrates the reliability of the instrument [53].

Also, as described, a second survey was performed to know the perceptions of the main technical leaders of *EOLO* regarding the process followed by the students and its impact on the project, which complements the results posed in the prior study [15]. The survey had 13 questions (8 closed-ended and 5 open-ended), 5 closed-ended questions on a 5 points Likert scale (Q1–Q5), 2 questions in the range (1) Very poor to (5) Excellent (Q6–Q7), one in the range 0–5 (Q8), which evaluated the overall performance of the students in *EOLO*, and the other five of the open-ended type. In some cases, given the number of questions or respondents, it was not possible to calculate the value of (α). The questions in the surveys respond to the EOs indicated in Section 3.3 and the need to know the perceptions of both the students and the technical staff of *EOLO* regarding the process and the methodology addressed. Although the instruments are self-reported, these provide us with important feedback about the methodology, the positive points, and aspects to improve for further methodologies and/or educational projects.

3.8. Analysis

To analyze the educational data, a mixed approach was adopted through embedded design, which is employed in engineering education studies [54]. In this, the quantitative and qualitative data collected with the mentioned instruments are analyzed in parallel to find the outcomes of the study and contrast them with the theoretical concepts or constructs. For the analysis of quantitative data of the surveys, e.g., mean (M) and standard deviation (SD), the software IBM SPSS v.27 was used, while the Computer Assisted/Aided Qualitative Data Analysis Software (CAQDAS) NVIVO v.12 was employed for the collecting and analysis of qualitative data thereof. Complementary, the main technical results of some projects are exposed, which include the lighting system, the CAN-Bus communication, the HMI, and the PV solar system.

4. Results and Discussion

4.1. Educational Results

Table 3 describes the descriptive statistics for the closed-ended questions in the student survey. Similarly, a chart with the distribution of the questions (Q1–Q8) is depicted in Figure 7.

Here, for example, 75% of the students considered that the different tasks performed in *EOLO* allowed the development of knowledge and learning (see Q3). This learning will be used by 58.3% of the students in their jobs (see Q5). Some of our students currently work in other fields of electronics different from the EVs and this could be a reason for the result. Despite the results in Q4 ($M = 3.83$, $SD = 1.4$), 75% of the students indicated that their motivation in the projects increased through the methodology, while 83.33% of the students in (Q1) employed the technical skills that were developed throughout their educational process in the different tasks and technical requirements in *EOLO*. In the same line, the students established that the teamwork helped the fulfillment of the different purposes and goals in each proposal, (see Q6: $M = 4.33$, $SD = 0.65$), which corroborates the coherence between the methodology and the collaborative learning. Even, the students manifested that the methodology and projects foster entrepreneurship, motivation, and teamwork as well as their problem-solving skills.

The students mentioned that they learned and applied concepts in the areas of programming, embedded systems, PCB design, battery measurement and control, design of 3D models in AutoCAD and CorelDraw, and photovoltaic systems, among others. These results are aligned with prior studies related to ESD and PBL. For instance, in the courses that employed this combination, students manifested an enhancement of their learning process and the possibility to apply their knowledge in relevant real-world problems [14,18,55]. This is also aligned with the results obtained in the prior work with $EOLO\ V_1$ [15]. For instance, in the survey followed in this prior study, the students classified the level of difficulty of each project between difficult (50%) and moderate (50%), and 58.3% of the students established that the generated apprenticeship in the project will serve in their future labor life. Similarly, 83.33% of the students employed the technical skills developed throughout their educational process.

Table 3. Descriptive statistics for the closed-ended questions in the student survey, ($n = 12$).

Question (Q)	M	SD
Q1. With the methodology used, Did you use in the design and implementation of your project, the knowledge and/or technical skills developed during your learning process?	4.17	1.19
Q2. Do you consider that the project developed was clear in its formulation and objectives?	3.92	1.16
Q3. Do you think that the methodology contributed to the development of your learning?	3.58	1.44
Q4. During the project, was your motivation increased to create the solution to the given problem?	3.83	1.4
Q5. Do you think that the learning in the development of your project will be used in your professional practice?	3.58	1.16
Q6. Do you think that the methodology allowed teamwork and collaboration, which helped to achieve the project's goals?	4.33	0.65
Q7. Were the tutors available to solve the different doubts and issues in your project?	4.5	0.67
Q8. Was it difficult to obtain the materials and equipment required for the design, development, and implementation of the solution to the problem posed in the *EOLO* project?	3.58	1.44
Q9. Please, assess your motivation and willingness in the developed activities and tasks in *EOLO*	4.42	0.51
Q10. Please, assess your performance working in team in *EOLO*	4.42	0.51
Q11. Please, evaluate the support of the tutors in relation to the *EOLO* project	4.75	0.45
Q12. What level of complexity do you consider the project developed at EOLO to have had?	3.5	0.52
Q13. Finally, taking into account the previous elements, please, assess the methodology in a range of 1–5.	4.33	0.78

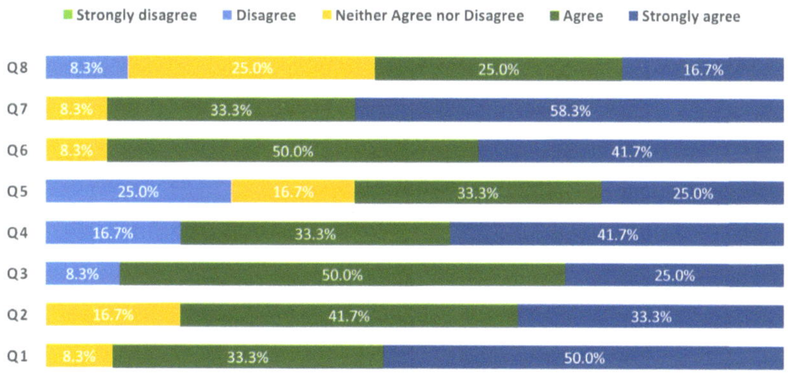

Figure 7. Distribution of the questions (Q1–Q8) for the students' survey on a Likert scale.

In Q8, the students indicated the difficulty to obtain materials and equipment. The result reveals a problem because some hardware devices were imported, increasing the time of development in each project. Thus, we tried to use hardware devices that were found in the local stores while the selected components arrived at each project. However, several students stated that it would have been a good idea to provide them with all the necessary materials for their projects. Under this issue, the original purpose in some projects was changed to meet the time requirements of the project, and the students evaluated this issue with the result of Q2 ($M = 3.92$, $SD = 1.16$). In this respect, some students (S3, S4) indicated some aspects to improve:

> *S3. Establish in advance the materials and products required since it takes a long time for these materials to be delivered and used for the development of the project.*

> *S4. Initially give a concrete and specific formulation of the project, making clear the work to be developed and its scope.*

Therefore, as an improvement action, it is necessary to reduce the time required for the purchase of materials and the administrative procedures at the university.

The students classified the level of difficulty of each project between difficult (50%) and moderate (50%), which means that the problems had a correct level of complexity and relevance in the engineering field (see Q12). This result is expected in methodologies that employ PBL since a requirement is the level of complexity and relevance of the projects created by the students [47,48].

Unfortunately, not all the tutors were available to the students in the project for time reasons. This fact could have affected the perception of other questions in the survey. Hence, this was a difficulty in the methodology and the students mentioned and evaluated this trouble in the survey. In this way, the tutor's accompaniment represented a critical point in the development of *EOLO*. From our perception, we believe it is important to improve the communication channels with the students. In this respect, two students (S1, S2) commented:

> *S1. I think that is a project with a vision and it has great potential, however, there is a lack of support from one tutor in the project. But also, I consider that the methodology functions well and it provides the opportunity for self-learning.*

> *S2. It is needed to assign tutors who are engaged with the project, otherwise, the execution time for the project can be affected.*

Nonetheless, to solve this issue, some tutors helped with the different doubts and inquiries of the students regardless of the time assigned for the project. For this reason, about 91.6% of the students in Q11 ($M = 4.75$, $SD = 0.45$) evaluated the support of the tutors between good and excellent. Concerning this, two students (S3, S4) mentioned:

> *S3. There was great support from the tutors and engineers of EOLO.*

> *S4. EOLO is an excellent project to learn and it has a broad field of knowledge application to explore and research.*

Despite the mentioned problems, the methodology has a good acceptance among the students ($M = 4.33$, $SD = 0.78$). Most of the students indicated that the methodology gave them the possibility to exchange knowledge and experiences with engineers and teammates, allowing to enhance their learning process.

As for the *EOLO* Engineers, Table 4 describes the results for the closed-ended questions in the survey applied to them. In this case, the SD was not calculated for each question due to the number of respondents. So, the engineers mentioned the following outstanding aspects of the students in the methodology:

- Interest to be participants of the *EOLO* Project.
- Compromise with their responsibilities.
- Self-demanding with their activities.
- Curiosity in all projects at the mechanical and electrical levels.

- Cooperation and disposition towards the activities outside their projects.

Table 4. Descriptive statistics for the *EOLO* engineer's survey, ($n = 2$).

Question (Q)	M
Q1. Do you think that the problems and solutions proposed in *EOLO* contributed to student learning?	5.0
Q2. Did the students accept the advice and feedback suggested to obtain the solution to the problems posed?	5.0
Q3. Do you think the students performed a correct implementation of the problem solution?	4.0
Q4. Do you think that the different learning obtained in the development of the projects by the students in *EOLO* will be used in their professional practice?	5.0
Q5. Do you believe that the methodology used in *EOLO* allowed teamwork that helped to achieve the project's goals?	5
Q6. Please evaluate the performance of students working as a team on the *EOLO* project.	4.5
Q7. Please evaluate the motivation of the students in the development of the project in *EOLO*.	4.5
Q8. Taking into account the above elements, on a scale of 1 to 5, evaluate the methodology followed by the students in the *EOLO* project.	4.5

The same engineers (E1, E2) evaluated the performance of the students in a range of (0) to (5) with an overall value of 4.5 (see Q8). As difficulties, recommendations, and comments, they indicated:

Difficulties:

E1. The students' time availability for the development of their activities and the socialization meetings.

E2. One difficulty may be to adjust the students' time with their advising teachers and the EOLO team. Also, due to the time constraints of the students, it is sometimes difficult for them to attend all the meetings proposed to socialize the planning of the projects as well as the progress, which directly affects the schedule of activities.

Recommendations:

E1. (1). To be able to generate more and better communication channels between students, tutors, and engineers of the EOLO project to achieve better control of developed activities and to coordinate more effectively the fulfillment of the activities. (2). More time available from the university so that tutors can give the necessary attention to their students. (3). More Documentation of progress made by the students in their respective projects.

E2. One recommendation may be to adjust the purchasing processes for materials, supplies, and equipment required for the project since the deadlines are very short and a delay in the purchasing process delays the projects.

Comments:

E1. At a general level, it has always been proposed that the university should bring its students closer to the industry and the EOLO project has been a research niche to involve students in industrial issues and more specifically in electric mobility, which is one of the new technologies that are beginning to position themselves worldwide.

E2. The problems or challenges faced by the students are based on real requirements, where they were able to demonstrate their capabilities, providing solutions tailored to the technical needs.

Previous comments and outcomes illustrate that is possible to create meaningful projects that enhance the learning process of the students through industry-academia collaboration. Nevertheless, it is necessary to improve the communication channels and reduce the administrative and purchasing times that affect the execution of the projects. As described, the students will employ the knowledge and learning created in the project for their labor life, which is an important consideration in the methodology with PjBL. With the industry–academia collaboration, the learning can be more situated to the real-world problems, in this case, in the area of renewable and sustainable energy, and EVs, contributing from the technical and educational perspectives to the achievement of the SDGs. The students directly put their knowledge to the advance of their projects with different hands-on activities, which is a clear differentiating factor in active learning methodologies. There are numerous success studies in the engineering context, mainly in software engineering, that illustrate the advantages that this type of collaboration (industry-academia) has on students [56–58], but also the challenges, in special with the relevance of the projects, the lack or drop of interest, the communication issues, and the monitoring process [59,60] that we have also experienced. Moreover, the industry–academia alliance can enable the development of high-quality personnel and motivate researchers to grow professionally within organizations [61].

Under these elements, our methodology integrated the components depicted in Figure 8. These elements can help to create meaningful learning with concrete and situated experiences that are included in the engineering curriculum. Finally, the previous results evidence that the OEs posed were accomplished and benefited the educational process of the students.

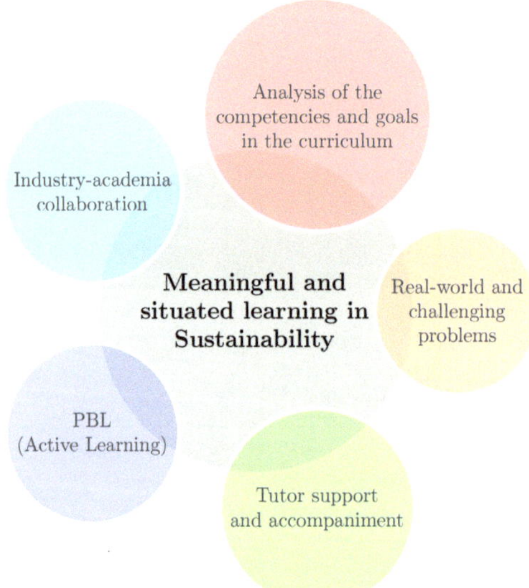

Figure 8. Main factors and components developed in the methodology.

4.2. Technical Results

4.2.1. Overall Results

At the technical level, the most representative results for the systems developed by the students were:

1. Reduction of wire gauge from AWG (14) to AWG (18), which decreases the costs relating to the electrical wiring.
2. Robustness in the vehicle communication through CAN-Bus protocol.
3. Incorporation of a Real-Time Operating System (RTOS) to prevent undesired delays in the functions of the hand-held control and in the screens commands.
4. Reduction of 72.3% in the power consumption in contrast with the first electric system in $EOLO\ V_1$. The first system had a power consumption of 318.5 W and the new system 88.2 W.

5. Low-cost user interface that gathers both a tachometer screen with vehicle variables such as speed, voltage, and current, and a second touch screen that provides the control of functions, improving the visual experience of the user.

In particular, the electric power reduction was analyzed with its measurement for each one of the subsystems made by the students vs. the previous one before their intervention as Table 5 shows. In this case, the user interface is composed of the touch and tachometer screens for the driver as Figure 6g depicts.

In *EOLO* V_1 shown in Figure 1a,b, the lighting system was composed of halogen lamps type (H4) [62] with a total current consumption of 10.67 A. These lamps were replaced with LED lamps whose maximum current consumption was 2.4 A. Additionally to this improvement, we increased the DC bus voltage in the lighting system from 12 V to 24 V through a 600 W DC-DC boost converter, searching for a reduction in the current of the LED lamps. These changes allowed a reduction of approximately 72% in the power consumption of the lighting system in comparison with the previous electric system installed in *EOLO*. In the same way, the initial user interface solely consisted of some signal lights installed on the steering wheel; therefore, an interface with screens was suggested to make the user experience more comfortable.

Table 5. Comparison of power consumption of the new electric system developed by the students vs. the previous one ($V_{battery}$ = 12.7 V). The nodes are those described in Figure 2.

Component/Subsystem	Power Consumption (First System)	Power Consumption (New System)
High Lights	129.1 W	30.6 W
Medium Lights	-	18.9 W
Low Lights	88 W	12.4 W
Turn Signals	18.9 W	5.5 W
Stationary Lights	32.7 W	8.7 W
Brake Light	49.9 W	2.4 W
Driver Interface	-	3.43 W
Nodes	-	6.4 W
Total	318.5 W	88.2 W

Some main critical requirements in the project were the reduction of the power consumption in the different nodes created in Figure 2, and the robustness of the CAN-Bus protocol under different events such as voltage transients, malfunction, and inadequate usage or handling by the driver. For this, we selected a 16-bit architecture with the devices (dsPIC33FJ128GP802) [40] that have an Enhanced CAN peripheral (ECAN) module incorporated for CAN-Bus communication [40,63], which have been used in prior projects [15,45,64] with feasible and good results from the educational and technical standpoints. The dsPICs provide advantages such as easiness of programming, tools for compiling and debugging such as MPLABX IDE, XC16, and ICD3, Direct Memory Access (DMA) peripheral in order to transfer data of the ECAN directly to RAM, and a maximum processor speed of 40 Million of Instructions per Second (MIPS). Moreover, the dsPICs have an architecture similar to 8-bit microcontrollers that are used in our curricula in the area of embedded systems, which benefited the learning of the students and accelerated the stage of design, development, and implementation. Some of the results for the systems implemented in the EV will be discussed below. It is worth mentioning that the architecture of the communication was created based on the prior works [65–67]. These works describe architectures with a CAN-Bus protocol to monitor variables in hybrid or cargo vehicles.

4.2.2. Front and Rear Lights (Lighting System)

Three LED lamps by side (left and right) were used with a maximum power consumption of 25 W. Each lamp has four power LEDs with its respective LED array for signaling. The intensity of each lamp is controlled by the device FL7760BM6X [43], which is a low-cost linear driver with a hysteresis current controller and a wide input range (8 V to 70 V DC). The FL7760BM6X uses a DC-DC Buck converter topology and it counts with thermal shutdown (TSD) and under-voltage lockout (UVLO) protections. The device receives a PWM signal from dsPIC on the (DIM) pin, which allows it to adjust the duty cycle of the signal to the MOSFET transistor (IRF740AS-Q1) whose operation frequency is 18 KHz. The control operates with the voltage regulation (200 mV $\pm$ 5%) over the resistor R_{senH}. Figure 9 shows the implemented schematic in the node. The values of the different components were taken by recommendation of the FL7760BM6X manufacturer [43], and adapted to the requirements of the lighting system according to Figure 2.

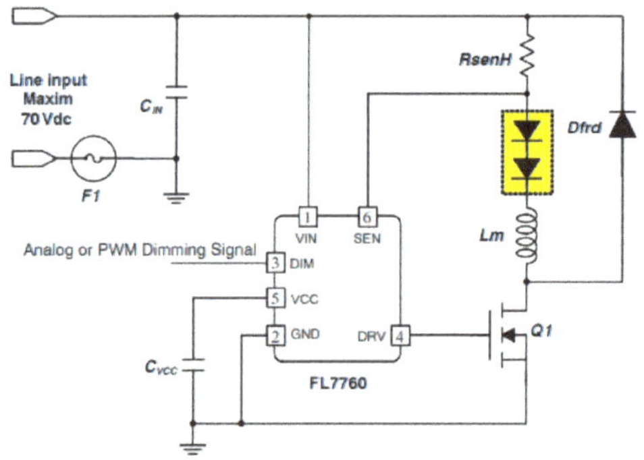

Figure 9. Implemented schematic for FL7760BM6X. R_{senH} = 0.11 Ω, L_m = 100 µH, C_{IN} = 470 µF/63 V, Q1 (IRF740AS), D_{frd} (B520C-Schottky diode).

Regarding the PWM frequency of 18 KHz for the transistor (Q1), this is configured by the Equation (1).

$$Timer_{Reg} = \frac{F_{CY}}{Pre \cdot F_{PWM}} \cdot k \tag{1}$$

where F_{CY} = 36.85 MHz, k is the duty cycle ($0 \leq k \leq 1$), ($pre = 1$) is the Timer pre-scale in the PWM peripheral, and F_{pwm} = 18 KHz. The frequency F_{CY} is obtained directly from the Internal Fast RC oscillator (FRC) of the dsPIC with the phase-locked loop (PLL) feature, providing an accuracy of $\pm$12%. For instance, whether ($k = 0.5$), these values yield to $Timer_{Reg}$ = 1024. Taking into account these elements, each one of the $Timer_{Reg}$ values for the front light functions (high, medium, low, and strobe) were characterized as Table 6 shows.

Table 6. Experimental duty cycle (k) for the front lights. The highest $Timer_{Reg}$ value is 2048 for $k = 1$.

Light Function	$Timer_{Reg}$ **Value**	**Duty Cycle (k)**
High	1300	0.635
Medium	800	0.4
Low	300	0.147
Strobe	(0–1300)	(0–0.635)

The different tests made with the lighting system pointed out a critical value for ($k = 0.8$) before a thermal failure in (Q1). Therefore, we limited the duty cycle to 0.635 preventing an over-temperature and premature damage in this transistor. The same scheme was followed for the rear lights with the difference that these lights had a current consumption of approximately $I_L = 100$ mA, which allowed us to use an array of BJT transistors compatible with automotive transients whose nominal collector current was $I_C = 500$ mA instead of the FL7760BM6X device. The appearance of the final PCB for the front lights installed in the vehicle is shown in Figure 10.

Figure 10. Example of the PCB for the front lights node. (1) Connector 24V. (2) dsPIC33FJ128GP802. (3) MCP2551 (CAN-Bus transceiver). (4) CAN-Bus connector. (5) Left lights connector. (6) FL7760BM6X (PWM controller). (7) Turn signals connector. (8) Right lights connector. (9) MOSFET transistor-*Q1* (IRF740AS).

4.2.3. Electronic Control Unit (ECU) Node

The ECU node processes and sends different commands for each node in the vehicle. Due to the large number of incoming events for the ECU node, we decided to incorporate a small Real-Time Operating System (RTOS) known as OSA [68]. OSA is a cooperative multitasking RTOS that only occupies around 4% of the dsPIC memory flash whose capacity is 128KB and it allows the designer to select and configure the timer for the different scheduling events, tasks, or semaphores. The tasks for RTOS are configured with their own priority (0–7) where (0) is the highest priority and so on. Two tasks (t_1, t_2) were set up in the project with a priority of (0) and (2), respectively. The difference between the priorities is because t_1 monitors the CAN-Bus communication while t_2 is for the handheld control in the steering of the EV.

Task (t_1) sends the light intensity and turn signal functions for the nodes that reply with a 29-bit identifier through a polling method with a monitoring period of 2 s. Whether this identifier matches the address allocated in the ECU node, a new command for the node is sent, otherwise, an error is generated. Task (t_2) checks the hand-held control and saves the function to be sent to the nodes. The hand-held control is interfaced directly with several General Purpose Input-Output (GPIOs) in the dsPIC with their respective pull-up resistors and RC filters. By the same token, the incoming events from the touch screen are processed by a UART interrupt that saves the information in the ECAN transmission buffer. At the same time, the UART interrupt sends the command to the tachometer screen through another available UART for further processing. In both cases, the UART peripherals have a baud rate of 9600 bps, which reduces data corruption produced by the noisy environment in the electrical system of the vehicle. The RTOS monitors every task (t_1, t_2) through a *time slice* of 1 ms provided by the Timer 1 of the dsPIC. A summarized scheme for the ECU functions is depicted in Figure 11.

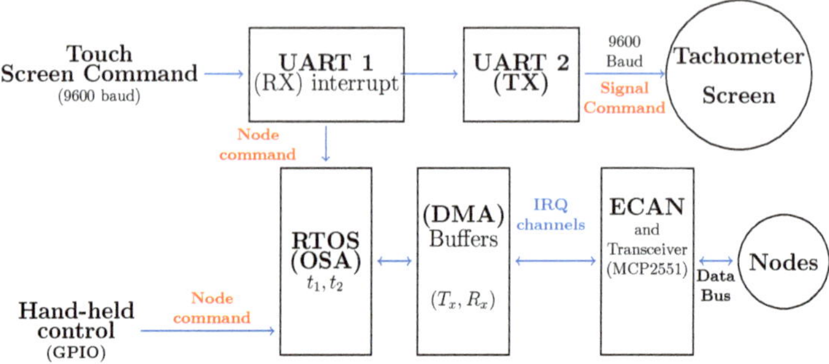

Figure 11. Overall embedded ECU components.

4.2.4. CAN-Bus Protocol

Each node obtains and processes a command from ECU by a polling method, which means that each node waits for a command in time intervals to prevent collision in the data bus of the CAN communication. A simplified Finite State Machine (FSM) diagram for the front lights node is depicted in Figure 12. A similar FSM diagram of Figure 12 for the front lights was followed for the rear lights with the functions of Figure 2. For readers interesting in the ECAN features of the dsPIC, please check the datasheet [40]. The communication sequence finally developed and implemented with CAN-Bus protocol for each node of the lighting system is described in the following steps:

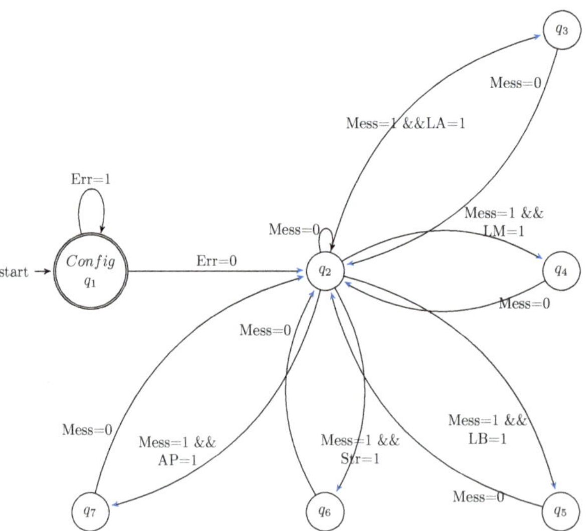

Figure 12. FSM diagram for the front lighting node.

- When a user starts the system (turn on the EV), state (q_1), the ECAN peripheral is configured with a default rate of 250 Kbps and a 29-bit identifier (CAN-2.0B) [63]. Additionally, the PWM is started with the mentioned frequency values. The CAN-Bus communication is type broadcast and solely the filters and masks provided in each node can select the adequate message for the node. Although the dsPIC has 15 filters and 3 masks, we decided to use filters (0-1) and masks (0-1) due to the small number of nodes in the communication. Whether no error is detected ($Err = 0$), the FSM passes the control to the state (q_2). Otherwise, the error is reported to the user ($Err = 1$). Notice that also the channels for Interrupt Request (IRQ) for the DMA peripheral must be configured with the appropriate values to manage the ECAN messages. The DMA directly transfers the information to the RAM buffers, which in this case are two, the first one for transmission and the other for reception. The size of each one is [1]*[8] (column vector).

- In the state (q_2), the node waits for an ECU message which contains the commands for the lights and signals (*Mess* = 0). In the case of a new message (*Mess* = 1), depending on the command, the FSM could pass to the states (q_3), (q_4), (q_5), (q_6), (q_7) for the light functions (high (LA), medium (LM), low (LB), strobe (Str) and turn off (AP)), respectively. The algorithm remains in its state until another command is received. A similar scheme presented in Figure 12 applies to the turn signals and stationary lights.
- A received command from ECU is a 16-bit concatenate value. That is, a single command contains the functions of the light intensity and the turn signals. An example of this is illustrated in Figure 13. The level of brightness in each level for the front and rear lights is according to the normative in the automotive sector in our country, which agrees with the duty cycle values shown previously in Table 6.

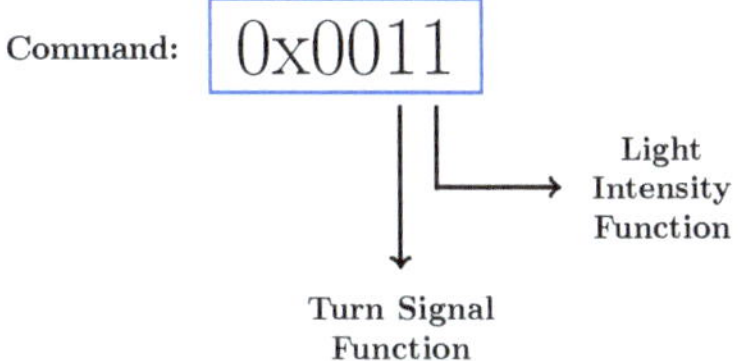

Figure 13. Example of concatenated 16-bit command from ECU.

The previous structure for any command prevents a collision due to the CAN-Bus being used once by the ECU. Tables 7 and 8 show the commands for light intensity and turn signals according to the structure of Figure 13. The commands can be selected either by the handheld control or the touch screen.

Table 7. Light intensity commands.

Command	Function
0	Turn off
1	High
2	Medium
3	Low
4	Strobe

Table 8. Turn signal commands.

Command	Function
0	Turn off
1	Turn Right
2	Turn Left
3	Stationary lights

In order to test the CAN-Bus protocol in the different nodes, we followed these steps:

1. CAN protocol communication (no functions): The purpose of this test was to check the polling method from the ECU node without functions. The polling method calls every node in Figure 2 and waits for a response. In case of no response, the system generates an alert for the driver. Then, we observed the possible transmission errors. Under this methodology, 30 controlled tests were made and 10 of them (30%) generated an error due to both the reduced time for the polling method and the electrical noisy environment. Our solution was to increase the time of polling in the RTOS and the processing time in each node gradually until the communication was stable. In the same way, we used a shielded cable wire in the vehicle for the CAN-Bus communication.
2. CAN protocol communication (with functions): In this test, we checked the functions of the hand-held control and the touch screen with the corrections made in the previous step. A set of functions were sent in short time intervals (10 s) from hand/held control and touch screen, searching for a malfunction in the RTOS or the CAN-Bus protocol. We performed 30 controlled tests without errors.
3. Start-up malfunction: In this test, we turned on and off the electrical system several times, searching for errors produced by the voltage transients in the start-up of the nodes. We found that the components more sensible to transients were the screens. Therefore, we designed and implemented a start-up mechanism with a microcontroller and a MOSFET transistor acting as a load switch. The mechanism let the filter bank capacitors in the battery charge for 3 s, and next the electrical system for the nodes was energized. Moreover, the nodes have a start-up period of 2 s.

4.2.5. Driver-focused Human Machine Interface (HMI)

The driver user interface was developed using two screens (Mikroelektronika HMI 7″) [42] that count with the features indicated in Table 9.

Table 9. Technical features of the screens used in the development of the user interface.

Feature	Description
Size	7″
Resolution	800 ∗ 480 px
Graphic Controller	FT812/13
Main Microcontroller	FT900Q (32-bit)
Microcontroller's speed	100 MHz/310 DMIPS
Flash Memory Capacity	256KB
MCU peripherals	CAN (2), SPI (2), I2C (2), I2S (1), UART (2)
Operating Voltage	3.3 V
Manufacturer	Mikroelektronica/Riverdi

We selected these screens because they offer high-quality graphics (16.7 M color depth and 400 NIT brightness), low current consumption (around 0.27 A for both screens), and widget support in the compiler toolkit. Regarding the tachometer screen, the speed is updated through a 12 V square signal derived from the electric engine and shifted with optocouplers to the 3.3 V level. The signal changes its frequency (f_t) from 4.62 Hz up to 30 Hz for a speed between ($\frac{10 \text{ km}}{h}$) and ($\frac{100 \text{ km}}{h}$). The vehicle's speed was modeled with a Curtis 1313 controller [69] that allowed it to change the motor's RPMs from its console taking into account the gear transmission ratio of the EV. This process yields the expression (2):

$$Speed(\frac{\text{km}}{\text{h}}) = 3.5461 \cdot f_t - 6.382 \tag{2}$$

The signal is captured by the input compare module in the dsPIC of the ECU that increases an internal counter/timer relying on the detected rising edges of this signal. When a period of 200 ms has concluded, the counter is checked and converted in frequency to apply Equation (2). Then, the speed value is sent to a tachometer widget.

Each widget or graphical component in the screen has an associated algorithm for the events and a driver. In the driver section, the attributes, I/O pins, and basic shapes that interact with the graphical controller (FT812) are specified. The events section contains the event-driven code when a widget, for example, is touched or clicked and it also has the possibility to interact with General Purpose I/O (GPIOs) and peripherals of the screens. The widgets described in this work were created through the tool Visual TFT [70]. The examples of the tachometer and touch screens are illustrated in Figure 14.

Figure 14. Example of driver-focused human interface (HMI). (**Top**) Tests of the HMI in the laboratory. (**Bottom**) HMI system installed on the EV.

Concerning the touch screen, we designed five sub-screens that have the functions of the main menu, time configuration, light intensity, conditioning air, and CAN-Bus communication test. A Real Time Clock (RTC) with I2C protocol (DS3231) installed on the touch screen allows the time configuration according to the user requirements. When a user changes the light intensity, a UART command is sent to the ECU node for its processing according to Figure 11. The CAN-Bus test command makes a routine to check the communication between nodes, sending the different functions for the vehicle. Both screens contain a security time of 2 s to start up the graphical components and to avoid malfunctioning in them.

4.2.6. Tests of Photovoltaic (PV) Panels

The utility with the described PV panels was to provide electric power to a parallel project concerning the air-conditioning of the EV. The aim of the PV panels is to provide the necessary power to this unit in the vehicle. For that, The PV panels were interfaced to a DC motor of the air-conditioning with three speeds and an operating voltage of 12 V. The DC motor creates an air circulation that cools hot air through a component called *evaporator*. Air-conditioning is custom and provided by a local manufacturer. Still, this concept is under test and the results exposed are a type of proof-of-concept of the system.

We measured the current and voltage of the DC motor + evaporator for the three speeds as Table 10 shows. The schematic with the connections for the PV panels, 12 V battery, and the evaporator is shown in Figure 15 with a low-cost solar PV panel controller of 20 A. In the same procedure, we identified the start-up current of the motor at these speeds. Unfortunately, the PV panels do not provide this current by themselves and this is the reason because we incorporated a 12 V–5 Ah lead–acid battery. The tests were made as Figure 6e depicts.

Considering these aspects, we performed some tests on a normal day to evaluate the feasibility of the system. The results are condensed in Table 11. The test was performed at Giradot, a Colombian town located at (74.7799 longitude and 4.3828 latitude). The average solar radiation for the 3 years in this town is 4.72 kWh/m, and the average ambient temperature is 23.06 °C [71]. This town has high rates of irradiance during the year.

In the data of Table 11, there are two elements to notice. The first one is that some values have a negative sign. In this case, this sign indicates that the device (battery or PV panels) is providing current to the air-conditioning. Otherwise, the positive sign indicates that the device (battery) is absorbing the current of the PV panels. The second one is that at certain hours (12:45 p.m.,12:55 p.m.) the maximum currents provided by the PV panels take place, and even in some cases the battery absorbs the current of the PV panels (see speeds 1 and 2). During these hours, the sky was clear and the day was fully sunny. These data point out that sustainable air-conditioning is feasible, and solar panels can generate a maximum of 88.83 W. However, this value must be improved for further developments, especially for speed 3, deeming also aspects such as the temperature of the driver's cabin, or solar PV panels with better efficiency. These values suppose an approximately 0.6% of increase in the efficiency taking into account the reported consumption of 14,529.9 W of the EV in the tests made in the document [31].

Table 10. Power consumption for the DC motor and evaporator in the air-conditioning unit.

Speed	Voltage (V)	Current (A)	Start-Up Current (A)	Power (W)
1	12.1	2.85	3.38	34.485
2	12.04	4.35	5.18	52.374
3	11.96	6.6	9.1	78.936

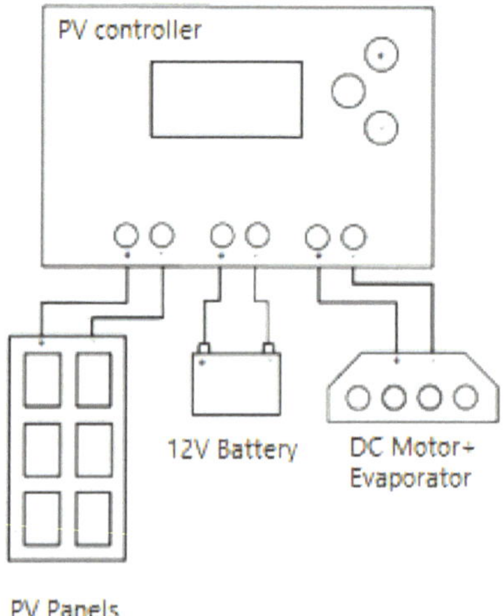

Figure 15. Schematic for the connections of the PV panels, battery, and evaporator (air-conditioning).

Table 11. Test of air-conditioning for the three available speeds. BC: Battery current (A), PV: Photo-voltaic panel current (A), SV: System voltage (V).

Speed	BC (A)	PV (A)	SV (V)	Time
	−0.1A	−3.22	12.7	11:45 a.m.
	1.48	−4.72	13	12:15 p.m.
1	2.77	−6.12	13.6	12:25 p.m.
	3.31	−6.65	14.4	**12:45 p.m.**
	2.66	−5.56	13.7	**12:55 p.m.**
	−1.61	−3.23	12.5	11:45 a.m.
	−0.08	−4.81	12.7	12:15 p.m.
2	1.12	−5.96	13	12:25 p.m.
	2.15	−6.73	13.2	**12:45 p.m.**
	2.21	−6.73	13.2	**12:55 p.m.**
	−3.72	−2.09	12.3	11:45 a.m.
	−2.47	−4.85	12.4	12:15 p.m.
3	−1.16	−5.95	12.6	12:25 p.m.
	0.1	−6.77	12.8	**12:45 p.m.**
	−0.54	−6.6	12.7	**12:55 p.m.**

5. Challenges and Recommendations

While the mentioned results show several advantages and success factors in the methodology, some challenges and issues have been detected and they should be taken into mind. A list of these and possible solutions based on the experience in this study are indicated as follows:

1. *Tutor accompaniment*: As described, the tutoring process was difficult in some cases due to the number of tasks and technical requirements in *EOLO*. Assigning more tutors for the projects and/or defining a smaller but relevant number of activities in each project in coordination with the technical staff in the industry can alleviate in part the issue. Tutor support and accompaniment are key factors in the methodology with PjBL even more when this last is carried out in an industry–academia collaboration.

2. *Reporting activities and follow-up*: Report progress achieved in shorter periods of time, expose advances, challenges, or issues under the peers and tutors, hold more meetings with the tutors, and use ICT tools for collaborative learning can be options explored to improve the report of the students' activities and their monitoring. Logbooks, short lab reports, and explanatory videos may be considered for the report.

3. *Administrative and purchasing times*: Explore open-source alternatives in hardware and software that could replace the current selection of devices or applications. Inquiry to local distributors for alternative devices or components. Use rapid prototyping techniques with development boards, 3D printers, laser cutters, or CNC machines [72–74].

4. *Sense of community*: Create a sense of community with a common goal in the project. All projects are important in the achievement of the objectives defined. The students are not working alone, but their designs, prototypes, or systems must interact with those created by their peers. Foster peer-to-peer collaboration. Create a healthy environment of collaboration and not competition. Share experiences with other colleagues (researchers–professors–faculty staff) even if they are of different knowledge areas. This can aid in obtaining different ideas and standpoints that could nourish the educational and technical objectives of the project.

5. *SDGs into the curricula*: Explore how SDGs and ESD can be incorporated under the lens of the curricula. Identify the competencies (soft and technical) that the students must develop and try to incorporate activities and active learning methodologies that foster SDGs. Clearly define learning objectives based on Bloom's or Marzano's taxonomies [75,76], and create, if it is possible, rubrics that help to evaluate the students. Think and create activities that engage the students regarding the SDGs.

6. *Communication channels*: Improve the communication channels between students, tutors, and technical staff involved in the project. Create regular meetings, brainstorming, mind map activities, and expose issues, challenges, or new ideas that will be convenient. Carefully consider students' views and opinions during the PjBL process.
7. *Complexity of the projects*: The level of complexity is fundamental to demand a high effort in the students, taking into account that this complexity must be attainable by achieving a solution in the projects. The above is based on the importance of reaching achievements and obtaining a solution to the problems proposed, which should have implications in the labor life of the students.
8. *Further projects*: Develop a bank of projects and possible problems in search of their solution, generating improvements to the devices and standardizing the requirements of the projects to be carried out.

6. Conclusions and Further Work

In this article, we described the educational aspects concerning the project *EOLO* with its implications in the process of learning, motivation, and collaboration of the students. Through an industry–academia collaboration, the students developed a set of projects that gave technical support to the different versions of *EOLO*. The deployed active learning methodology with PjBL allowed the enhancement of the learning process of the students and encouraged peer collaboration, motivation, and problem-solving skills. These results are aligned with prior works such as [14,18], where students, through discovery and inquiry, searched for solutions for diverse problems in the field of renewable energy. Nonetheless, in counterpart to prior studies which are situated from a holistic view from HEIs [9,13], we think that the initiatives from educators in classrooms are important to achieve and promote SD. With these initiatives, the change in SD can also be fostered in HEIs.

At the educational level, some points such as the tutor's accompaniment, the project's complexity, the reduction of the administrative and purchasing times, the follow-up and report of the performed activities, and the sense of community should be revisited to improve the methodology. Additionally, the standpoint of the technical staff of *EOLO* regarding the performance of the students indicates that the engagement, compromise, curiosity, and cooperation by the students are the most remarkable educational and attitudinal aspects of the methodology. In the same way as the studies [9,13,14,18], we think the most difficult issue in SD is its effective incorporation in the classrooms. Therefore, in this study, we constructed a methodology within a representative project in the renewable energy field. Perhaps, another challenging issue is finding a relevant project that engages students, and that it catalyzes learning, motivation, self-efficacy, etc. This is the main reason why an industry–academia collaboration is suitable to create and develop these types of projects. As aspects to improve, we think that is necessary to widen the scope of the methodology to other engineering and SD fields, reinforcing the previous weakness points found in the prior work [15] and in the current study. Although we believe that the incorporation of 12 students is appropriate considering the features of PjBL, the validation of the methodology with more students could lead us to better improvements and results from the educational standpoint.

At the technical level, this study described a system through a CAN-Bus protocol and a user interface developed with the support of the students that comply with the design requirements and robustness parameters. Moreover, we achieved, with the developed electric and communication architecture, a reduction of 72.3% in the power consumption in the lighting system in contrast with the first electric system deployed in *EOLO*. Also, some tests regarding the PV panels or the wind turbine suggest that an improvement in the EV's efficiency is possible. Nonetheless, further development and testing are necessary to fulfill the goal of the efficiency of 11%, and the robustness of the nodes in the EV. Still, several components, nodes, and renewable energy mechanisms are under test in *EOLO*.

All of these elements allow us to state that the methodology was successful in the objectives and goals proposed. However, the mentioned issues and recommendations should be carefully taken into account for further developments or educational designs of active learning methodologies. Further work will be focused on strengthening the weak points of the methodology and include other projects in the fields of renewable energy, the Internet of Things (IoT), machine learning, and AI in which new students will be involved in *EOLO*.

Author Contributions: Conceptualization , J.Á.A.; Methodology, J.Á.A.; Investigation, J.Á.A. and T.G.O.-A.; Resources, J.Á.A.; Writing—original draft, J.Á.A.; Writing—review & editing, J.Á.A. and T.G.O.-A.; Visualization, T.G.O.-A.; Supervision, T.G.O.-A. All authors have read and agreed to the published version of the manuscript.

Funding: This study was supported by the Corporación Universitaria Minutos de Dios-UNIMINUTO under grant CSP4-16-067.

Institutional Review Board Statement: Not applicable.

Informed Consent Statement: Not applicable.

Acknowledgments: The authors wish to extend their acknowledgment to the students that participated in the design, implementation, and debugging of the different nodes in the vehicle, without their efforts this project would have not been possible. Also, we extend an acknowledgment to Mauricio Roldán and Professor Heyson Báez for their support, advice, patience, and work in *EOLO*.

Conflicts of Interest: The authors declare no conflict of interest.

References

1. UNESCO. Sustainable Development Goals (SDGs). Available online: https://sdgs.un.org/goals (accessed on 3 April 2023).
2. UNESCO. The Sustainable Development Goals Report. 2022. Available online: https://unstats.un.org/sdgs/report/2022/The-Sustainable-Development-Goals-Report-2022.pdf (accessed on 3 April 2023).
3. UNESCO. Overview of Sustainable Development Goal 7. Available online: https://sdgs.un.org/goals/goal7 (accessed on 3 April 2023).
4. UNESCO. The 2030 Agenda for Sustainable Development. Available online: https://sustainabledevelopment.un.org/content/documents/21252030AgendaforSustainableDevelopmentweb.pdf (accessed on 3 April 2023).
5. UNESCO. Criteria for Accrediting Engineering Programs, 2020–2021. Available online: https://www.abet.org/accreditation/accreditation-criteria/criteria-for-accrediting-engineering-programs-2020-2021/ (accessed on 3 April 2023).
6. Leal Filho, W.; Pace, P. *Teaching Education for Sustainable Development at University Level*; Springer: Berlin/Heidelberg, Germany, 2016.
7. Universities, S. From Declarations on Sustainability in Higher Education to National Law by Thomas Skou Grindsted. *J. Environ. Econ. Manag.* **2011**, *2*, 29–36.
8. Wright, T. The evolution of sustainability declarations in higher education. In *Higher Education and the Challenge of Sustainability: Problematics, Promise, and Practice*; Springer: Berlin/Heidelberg, Germany, 2004; pp. 7–19.
9. Miñano Rubio, R.; Uribe, D.; Moreno-Romero, A.; Yáñez, S. Embedding sustainability competences into engineering education. The case of informatics engineering and industrial engineering degree programs at Spanish universities. *Sustainability* **2019**, *11*, 5832. [CrossRef]
10. Mulder, K.F.; Segalas, J.; Ferrer-Balas, D. How to educate engineers for/in sustainable development: Ten years of discussion, remaining challenges. *Int. J. Sustain. High. Educ.* **2012**, *13*, 211–218. [CrossRef]
11. Lozano, R.; Ceulemans, K.; Alonso-Almeida, M.; Huisingh, D.; Lozano, F.J.; Waas, T.; Lambrechts, W.; Lukman, R.; Hugé, J. A review of commitment and implementation of sustainable development in higher education: Results from a worldwide survey. *J. Clean. Prod.* **2015**, *108*, 1–18. [CrossRef]
12. Cebrián, G.; Palau, R.; Mogas, J. The smart classroom as a means to the development of ESD methodologies. *Sustainability* **2020**, *12*, 3010. [CrossRef]
13. Neto, V. Eco-design and Eco-efficiency Competencies Development in Engineering and Design Students. *Educ. Sci.* **2019**, *9*, 126. [CrossRef]
14. Ulazia, A.; Ibarra-Berastegi, G. Problem-based learning in university studies on renewable energies: Case of a laboratory windpump. *Sustainability* **2020**, *12*, 2495. [CrossRef]
15. Ariza, J.Á.; Ramos, H.B. An educational experience with PBL in capstone projects: The case of EOLO a Colombian electric vehicle with sustainable energy systems. In Proceedings of the 2019 IEEE Frontiers in Education Conference (FIE), Cincinati, OH, USA, 16–19 October 2019; pp. 1–9.
16. Block, B.M.; Haus, B. New ways in engineering education for a sustainable and smart future. In Proceedings of the 2020 IEEE Frontiers in Education Conference (FIE), Uppsala, Sweden, 21–24 October 2020; pp. 1–9.

17. Curiel-Ramirez, L.A.; Bautista-Montesano, R.; Galluzzi, R.; Izquierdo-Reyes, J.; Ramírez-Mendoza, R.A.; Bustamante-Bello, R. Smart automotive E-Mobility—A proposal for a new curricula for engineering education. *Educ. Sci.* **2022**, *12*, 316. [CrossRef]
18. Wang, X.; Guo, L. How to promote university students to innovative use renewable energy? An inquiry-based learning course model. *Sustainability* **2021**, *13*, 1418. [CrossRef]
19. Oliveira, J.; Neves, L.; Lanceros-Mendez, S. Kit "Energy, Environment and Sustainability": An educational strategy for a sustainable future. A case study for Guinea-Bissau. *Educ. Sci.* **2021**, *11*, 787. [CrossRef]
20. Forero-García, E.; Castañeda, D.P.; Corredor-Cely, J.; Paternina, J.L. Energetic Competencies in Electronic Engineering Education: A Sustainable Social Commitment. *J. Eng. Educ. Transform.* **2022**, *36*, 55–66. [CrossRef]
21. Rendón López, L.M.; Escobar Londoño, J.V.; Arango Ruiz, Á.D.J.; Molina Benítez, J.A.; Villamil Parodi, T.; Valencia Montaña, D.F. Educación para el desarrollo sostenible: Acercamientos desde una perspectiva colombiana. *Prod.+ Limpia* **2018**, *13*, 133–149. [CrossRef]
22. Lugo, L.H.V. La ingeniería en Colombia, ¿educación de calidad?: El cuarto Objetivo de Desarrollo Sostenible de la ONU. *Educ. Desarro. Soc.* **2018**, *12*, 60–73.
23. González, M.E.B.; Londoño, S.S.; Méndez, L.C.V.; Martina, M.A.M. Educación Para La Sostenibilidad En Ingeniería Ambiental Como Aporte Al Desarrollo Social. In Proceedings of the 2020 Encuentro Internacional de Educación en Ingeniería ACOFI, Bogotá, Colombia, 2 December 2020; pp. 1–8.
24. Ashford, N.A. Major challenges to engineering education for sustainable development: What has to change to make it creative, effective, and acceptable to the established disciplines? *Int. J. Sustain. High. Educ.* **2004**, *5*, 239–250. [CrossRef]
25. Denby, L.; Rickards, S. An approach to embedding sustainability into undergraduate curriculum: Macquarie university, Australia case study. In *Teaching Education for Sustainable Development at University Level*; Springer: Berlin/Heidelberg, Germany, 2016; pp. 9–33.
26. Glavič, P. Identifying key issues of education for sustainable development. *Sustainability* **2020**, *12*, 6500. [CrossRef]
27. Fathabadi, H. Utilizing solar and wind energy in plug-in hybrid electric vehicles. *Energy Convers. Manag.* **2018**, *156*, 317–328. [CrossRef]
28. Kassem, Y. Computational study on vertical axis wind turbine car: static study. *Model. Earth Syst. Environ.* **2018**, *4*, 1041–1057. [CrossRef]
29. Quartey, G.; Adzimah, S.K. Generation of electrical power by a wind turbine for charging moving electric cars. *J. Energy Technol. Policy* **2014**, *4*, 19.
30. Monzamodeth, R.S.A.; Nicolás Iván, R.R.; Oscar, X.; Bernardo, H.M.; Osvaldo, F.; Fermín, C.; Bernardo, C. Development of a wind turbine using 3D printing: A prospection of electric power generation from daily commute by car. *Wind. Eng.* **2022**, *46*, 376–391. [CrossRef]
31. Garcia, A.; Reyes, J.S.; Wang, X.; Roldan, J.; Olaya, M. Performance Study of an Electric Vehicle "Eolo" with a Mounted Aeolian Generator. In Proceedings of the ASME International Mechanical Engineering Congress and Exposition. American Society of Mechanical Engineers, Virtual Conference (Online), 1–5 November 2021; Volume 85642, p. V08BT08A037.
32. de Dios, C.I.M. CIMD Mission and Overview. Available online: https://mdc.org.co/ (accessed on 25 March 2023).
33. Abraham, J.; Plourde, B.; Mowry, G.; Minkowycz, W.; Sparrow, E. Summary of Savonius wind turbine development and future applications for small-scale power generation. *J. Renew. Sustain. Energy* **2012**, *4*, 042703. [CrossRef]
34. de Negocios, C.V. Carro Eléctrico: Proyecto EOLO en Colombia. Available online: https://cvn.com.co/admincvn/carro-electrico-proyecto-eolo-colombia/ (accessed on 25 March 2023).
35. AutoLab. Eolo: El Primer Carro Eólico Colombiano. Available online: https://autolab.com.co/blog/eolo-carro-eolico-colombiano/ (accessed on 25 March 2023).
36. Castro, N.J.R.; Pareja, M.L.I.; Dávila, M.O. Desarrollo de una máquina eólica de doble turbina de eje vertical para la generación de energía en un auto eléctrico. *Ingeniare* **2015**, *19*, 117–126. [CrossRef]
37. Álvarez Ariza, J. Technical Results of the Electric Vehicle EOLO. Available online: https://www.youtube.com/watch?v=tgbfEwdet8U (accessed on 27 March 2023).
38. Herazo, C.A.; Ariza, J.Á. A proposal of educational model for research incubators in technological programs of electronics. In Proceedings of the 2016 IEEE 8th International Conference on Engineering Education (ICEED), Kuala Lumpur, Malaysia, 7–8 December 2016; pp. 143–148.
39. gomog. Vehicle Wiring Colour Code. Available online: https://www.gomog.com/allmorgan/LucasColours.html (accessed on 25 April 2023).
40. Microchip Technology Inc. dsPIC33FJ32GP802 Datasheet. Available online: http://ww1.microchip.com/downloads/en/devicedoc/70292e.pdf (accessed on 2 February 2023).
41. Future Technology Devices International Limited (FTDI). FT900Q Microcontroller Datasheet. Available online: https://www.ftdichip.com/Support/Documents/DataSheets/ICs/DS_FT900_1_2_3.pdf (accessed on 2 February 2023).
42. Mikroelektronika. HMI 7 Technical Specifications. Available online: https://www.mikroe.com/mikromedia-hmi-70-no-touch (accessed on 2 February 2023).
43. Semiconductor Components Industries, LLC. FL7760 Datasheet. Available online: https://www.onsemi.com/pub/Collateral/FL7760-D.PDF (accessed on 2 February 2023).

44. Álvarez Ariza, J. Dscblocks: An open-source platform for learning embedded systems based on algorithm visualizations and digital signal controllers. *Electronics* **2019**, *8*, 228. [CrossRef]
45. Ariza, J.A. Controlly: Open source platform for learning and teaching control systems. In Proceedings of the 2015 IEEE 2nd Colombian Conference on Automatic Control (CCAC), Manizales, Colombia, 14–16 October 2015; pp. 1–6.
46. University of Nottingham. Xerte Authoring Tool Webpage. Available online: https://xerte.org.uk/index.php/en/ (accessed on 23 January 2023).
47. Aksela, M.; Haatainen, O. Project-based learning (PBL) in practise: Active teachers' views of its' advantages and challenges. In Proceedings of the 5th International STEM in Education Conference Proceedings: Integrated Education for the Real World, Brisbane, Australia, 21–23 November 2018.
48. Kokotsaki, D.; Menzies, V.; Wiggins, A. Project-based learning: A review of the literature. *Improv. Sch.* **2016**, *19*, 267–277. [CrossRef]
49. Seel, N.M. *Encyclopedia of the Sciences of Learning*; Springer Science & Business Media: Berlin/Heidelberg, Germany, 2011.
50. Dumitrache, A.; Gheorghe, M. Project based learning. Practical steps in completing a learning assignment. In Proceedings of the the International Scientific Conference eLearning and Software for Education, Bucharest, Romania, 19–20 April 2018; "Carol I" National Defence University: Bucharest, Romania, 2018; Volume 1, pp. 95–100.
51. Zhang, J.; Zhang, Z.; Philbin, S.P.; Huijser, H.; Wang, Q.; Jin, R. Toward next-generation engineering education: A case study of an engineering capstone project based on BIM technology in MEP systems. *Comput. Appl. Eng. Educ.* **2022**, *30*, 146–162. [CrossRef]
52. Friess, W.A.; Goupee, A.J. Using continuous peer evaluation in team-based engineering capstone projects: A case study. *IEEE D* **2020**, *63*, 82–87. [CrossRef]
53. Yu, C.H. An introduction to computing and interpreting Cronbach Coefficient Alpha in SAS. In *Proceedings of the 26th SAS User Group International Conference*; SAS Institute Inc.: Cary, NC, USA, 2001; Volume 2225, pp. 1–6.
54. Borrego, M.; Douglas, E.P.; Amelink, C.T. Quantitative, qualitative, and mixed research methods in engineering education. *J. Eng. Educ.* **2009**, *98*, 53–66. [CrossRef]
55. Belu, R.G.; Husanu, I.N.C. Embedding renewable energy and sustainability into the engineering technology curricula. In Proceedings of the 2012 ASEE Annual Conference & Exposition, San Antonio, TX, USA, 10–13 June 2012; pp. 25–518.
56. Hanieh, A.A.; AbdElall, S.; Krajnik, P.; Hasan, A. Industry-academia partnership for sustainable development in Palestine. *Procedia CIRP* **2015**, *26*, 109–114. [CrossRef]
57. Wohlin, C.; Runeson, P. Guiding the selection of research methodology in industry–academia collaboration in software engineering. *Inf. Softw. Technol.* **2021**, *140*, 106678. [CrossRef]
58. Marijan, D.; Gotlieb, A. Industry-Academia research collaboration in software engineering: The Certus model. *Inf. Softw. Technol.* **2021**, *132*, 106473. [CrossRef]
59. Garousi, V.; Felderer, M.; Fernandes, J.M.; Pfahl, D.; Mäntylä, M.V. Industry-academia collaborations in software engineering: An empirical analysis of challenges, patterns and anti-patterns in research projects. In Proceedings of the 21st International Conference on Evaluation and Assessment in Software Engineering, Karlskrona, Sweden, 15–16 June 2017; pp. 224–229.
60. Marijan, D.; Sen, S. Industry–academia research collaboration and knowledge co-creation: Patterns and anti-patterns. *ACM Trans. Softw. Eng. Methodol. (TOSEM)* **2022**, *31*, 1–52. [CrossRef]
61. Gandhi, M. Industry-academia collaboration in India: Recent initiatives, issues, challenges, opportunities and strategies. *Bus. Manag. Rev.* **2014**, *5*, 45.
62. HELLA, Inc. Features of Halogen Lamps Type H4. Available online: https://www.hella.com/hella-csa/assets/media_global/CSA_Bulbs_Catalogue_2018-2019_SP_LRes.pdf (accessed on 20 April 2023).
63. *Bosch Controller Area Network (CAN)*; Version 2.0; Bosch: Gerlingen, Germany, 1998; pp. 1–72.
64. Ariza, J.A. Design of open source platform for automatic control systems education based on cooperative learning. In Proceedings of the 2016 IEEE Frontiers in Education Conference (FIE), Erie, PA, USA, 12–15 October 2016; pp. 1–9.
65. Fernandes, J.N.O. A real-time embedded system for monitoring of cargo vehicles, using controller area network (CAN). *IEEE Lat. Am. Trans.* **2016**, *14*, 1086–1092. [CrossRef]
66. Gurram, S.K.; Conrad, J.M. Implementation of CAN bus in an autonomous all-terrain vehicle. In Proceedings of the 2011 Proceedings of IEEE Southeastcon, Nashville, TN, USA, 17–20 March 2011; pp. 250–254.
67. Livint, G.; Horga, V.; Ratoi, M.; Albu, M.; Chiriac, G. Implementing the CANopen protocol for the distributed control of a hybrid electric vehicle. In Proceedings of the 2009 8th International Symposium on Advanced Electromechanical Motion Systems & Electric Drives Joint Symposium, Lillie, France, 1–3 July 2009; pp. 1–6.
68. Timofeev, V. OSA RTOS. Available online: http://wiki.pic24.ru/doku.php/en/osa/ref/introduction/intro (accessed on 10 April 2023).
69. Curtis Instruments, I. Curtis 1313 Users' Manual. Available online: https://www.thunderstruck-ev.com/images/1313%20Manual.pdf (accessed on 17 April 2023).
70. Mikroelektronika. Visual TFT. Available online: https://www.mikroe.com/visual-tft (accessed on 7 April 2023).
71. Restrepo, S.A.; Morcillo, J.; Castaneda, M.; Zapata, S.; Aristizábal, A.J. Experimental research on the performance of a BIPV system operating in Girardot, Colombia. *Energy Rep.* **2023**, *9*, 194–204. [CrossRef]
72. Irwin, J.L.; Pearce, J.M.; Anzalone, G.C.; Douglas, M.; Oppliger, E. The RepRap 3-D printer revolution in STEM education. In Proceedings of the 2014 ASEE Annual Conference and Exposition, Indianapolis, IN, USA, 15–18 June 2014; pp. 1–13.

73. Pearce, J.M.; Anzalone, N.; Heldt, C. Open-source wax RepRap 3-D printer for rapid prototyping paper-based microfluidics. *SLAS Technol.* **2016**, *21*, 510–516. [CrossRef] [PubMed]
74. Laplume, A.O.; Petersen, B.; Pearce, J.M. Global value chains from a 3D printing perspective. *J. Int. Bus. Stud.* **2016**, *47*, 595–609. [CrossRef]
75. Krathwohl, D.R. A revision of Bloom's taxonomy: An overview. *Theory Pract.* **2002**, *41*, 212–218. [CrossRef]
76. Marzano, R.J.; Kendall, J.S. *The New Taxonomy of Educational Objectives*; Corwin Press: Thousand Oaks, CA, USA, 2006.

 *sustainability*

Article

Engineering Students' Perception on Self-Efficacy in Pre and Post Pandemic Phase

Clara Viegas *, Natércia Lima and Alexandra R. Costa

CIETI/ISEP, Polytechnic of Porto, R. Dr. António Bernardino de Almeida, 4249-015 Porto, Portugal; nmm@isep.ipp.pt (N.L.); map@isep.ipp.pt (A.R.C.)
* Correspondence: mcm@isep.ipp.pt; Tel.: +351-228-340-500

Abstract: During 2020 and 2021, the world experienced a global change in everyone's daily lives due to the COVID-19 pandemic. Students were confined in their homes but, luckily, had access to online classes. This study aims to assess the changes in self-efficacy perceived by engineering students in a school in Portugal. By helping to understand how students have changed their learning capacities, developed new strategies, and/or need more (or different) support to learn, teachers can target their teaching methods accordingly and contribute to a more sustainable education. A questionnaire was constructed and validated to assess students' perceptions before and after the associated lockdowns. Five theoretically supported factors emerged from a statistical factor analysis: Communication and Empathy; Focus and Personal Organization; Teamwork and Individual Work Capacity; Technical and Cognitive Resources Management; and Emotional Resources Management. This work shows students' percept that they improved their teamwork and individual work capacity and their technical and cognitive resources management. In general, students seem to have been able to be more autonomous as they managed to work and develop their cognitive resources; however, their emotional state and ability to focus decreased. Perceived self-efficacy was less affected in older students than in younger ones, suggesting that this group may have adapted better to the pandemic restrictions. Students who were already at university showed less impact than those moving from high school to university. There was also a difference between those who endured these changes at only one level of education and those who endured them at both levels (high school and university), with this last group being the most negatively affected.

Keywords: students' perception; learning; higher education; students' academic competences; students' interpersonal competences; self-efficacy

 check for updates

Citation: Viegas, C.; Lima, N.; Costa, A.R. Engineering Students' Perception on Self-Efficacy in Pre and Post Pandemic Phase. *Sustainability* **2023**, *15*, 9538. https://doi.org/10.3390/su15129538

Academic Editor: Grigorios L. Kyriakopoulos

Received: 24 April 2023
Revised: 24 May 2023
Accepted: 5 June 2023
Published: 14 June 2023

1. Introduction

Few moments in human history have brought such rapid and marked changes as the COVID-19 pandemic crisis. In the field of education, thousands of students and teachers around the world have been placed at home using online platforms as schools. The long periods of confinement that took place between 2020 and 2022 isolated students from their social contacts, locking them in their houses. Their rooms became, at the same time, a bedroom, a place of study, and their classroom; these were the most fortunate. Some students went through greater limitations having to share space and resources with others or even not having access to the resources they needed, such as computers or the internet [1,2].

Many researchers have studied the effects of these lockdowns on students, revealing consequences in terms of sociability, lack of motivation and interest in class and learning, and even in terms of mental illness, namely increased levels of anxiety, stress, and depression [3–5].

After three years, the pandemic seems to have subsided, and gradually students are returning to the classroom. This return to normality will not, however, erase the experiences lived during the pandemic, several months of confinement, and long periods of homeschooling. According to UNESCO, "In a post-COVID-19 world there will be a great need to cure the separations that have arisen due to quarantines and distancing restrictions. We will need to think creatively about ways to reconnect people. Trusting young people and empowering them to think and act together is one important way to accomplish this" [6] (p. 14). Lessons can be learned if we understand more clearly what the major changes were, some of them irreversible, and how to work upon this new scenario to build an even stronger education, perhaps using some strategies and resources developed during that time. Nor teachers nor students are the same after this predicament, so understanding what really changed in terms of students' learning might be a way to develop a more sustainable education.

Even though the pandemic period has been extensively studied in recent research, there are still some important issues that need our attention, namely how students perceive classes and their learning in this return to normality. Are students learning in the same way as before? Are they developing the necessary competences equally as before? Do students percept any change in his/her behavior or study habits? These are some of the questions that are not yet fully addressed in the literature and are fundamental to better understand students' reality. The problem addressed in this work aims to contribute to this better understanding of students' perceptions regarding this return to face-to-face teaching in terms of how the years in confinement have affected them. This way, teachers may better adapt to this new reality, adequate their teaching and methodologies or resources to enhance students' learning and develop the necessary competences.

This study was developed in a higher education institution, namely the school of engineering, with 487 participants through a survey. The primary goal of this work is to identify differences in students' behavior (compared to the prior pandemic era), namely interpersonal and academic competences. These competences may be influenced by students' perception of their self-efficacy. A secondary objective is to understand whether there are significant differences between students who undertook these restrictions mainly in high school and those who were already at university.

In addition to the introduction, this paper is organized into five sections and its aim is to study whether there are statistically significant differences between students' perceptions of their self-efficacy when comparing the pre- and post-pandemic phases. Section 2 is devoted to the effects of the pandemic on education, focusing on students and a theoretical background on students' perception of self-efficacy. In Section 3, the research methodology is lined up, and the reliability and validity of the study are addressed. Then, in Section 4, the results are presented and analyzed. Finally, Sections 5 and 6 discuss the results with the literature and draw conclusions that contribute to the research question: *"In which way the COVID-19 Pandemic with its lockdowns affected students' self-efficacy perception?"*.

2. State of the Art

2.1. Students' Classrooms since the Pandemic

Between 2020 and 2021, the world lived through the COVID-19 pandemic, and most countries endured global lockdowns that lasted for months. Schools and universities developed online teaching during those periods. During the pandemic, teachers and students adapted to online conditions and, in several cases, reinvented their way of teaching and their way of learning. Some studies [7–10] indicate that teachers made an effort to diversify teaching resources and applied new methodologies and support. Other studies [11–13] also indicate that students struggled with online classes and felt that they were more productive when there was interaction between the teacher and the students or between students. Some students deeply felt the lack of contact with teachers and other students [5,14].

After enduring nearly two years with restrictions to face-to-face classes, higher education institutions returned to their normal functioning. According to Hess [15], despite all the experiences and everything that happened during the pandemic period, the majority of schools returned to their usual rhythms and routines, and little was changed. Teachers and students, eager to return to the face-to-face mode, drop most of the online resources and materials developed during the pandemic [15]. Despite this absence of changes in terms of routines and in terms of school life, it is expected that such a long period of isolation will cause behavioral changes, particularly in young people, for whom social support from the group they belong to is of great importance. Therefore, it is important to study this return to face-to-face learning, namely by listening to students and understanding their perceptions of what has changed since the end of the pandemic period. Several research studies have documented how students reacted to this new reality during the pandemic, but few mention the modifications students felt in their academic and interpersonal competences. In fact, the researchers' focus on students' perception of the impact of the pandemic period focused on their opinions about the shift from face-to-face study to online study [11,16–18]. Little attention has been paid to the student's views on what has changed since their return to face-to-face learning, how they perceive the relationships with their colleagues and teachers, the productivity of their learning in general, and their emotional state after this period. One of the exceptions is the study developed by Becker et al. [19]. According to these authors, the impact of the long period of confinement on the students' skills may vary depending on the student's previous experience, that is, whether or not they were already attending the same degree prior to confinement [19]. This work is intended to contribute to the continued filling of this gap with the goal of addressing students' perceptions of themselves and their learning and how and what has changed.

2.2. Students' Perceptions

According to Curelaru et al. [3], "Perceptions are defined as complex mental processes by which people understand, interpret, evaluate, and form a picture of social phenomena" (p. 2). This interpretation of reality obviously affects the behavior of individuals. Thus, students' perceptions of how they were affected by the lockdowns may now, on their return to face-to-face classes, influence their behavior, attention, motivation, emotions, and satisfaction level.

Students' self-assessment is important as it represents a measure of self-efficacy that may influence students' behavior. Self-efficacy has been conceptualized as the belief that individuals have in their own ability to organize and execute the necessary actions to achieve certain goals [20]. Self-efficacy beliefs are, therefore, an important part of the motivational process, influencing the way the subject prepares for action. A positive belief in one's ability to perform a task can encourage behaviors that, by facilitating success, ultimately reinforce the belief in self-efficacy. Individuals with high levels of self-efficacy prefer to develop more challenging tasks and set more demanding goals for themselves; invest, at the same time, more in the tasks in which they are involved, showing greater levels of effort and persistence, overcoming more quickly the difficulties they face and maintaining focus on defined objectives [20–23]. Students' positive beliefs about their self-efficacy to manage academic tasks may also emotionally influence them by decreasing stress, anxiety, and depression [24]. By assessing students' self-efficacy beliefs, one can infer information about their predisposition to engage, invest and persist in learning activities.

Engineering students need to address several academic (subject-related) and interpersonal competences (soft skills) to fully cope with the profession when they graduate [25–27]. According to Alison Doyle [28], interpersonal skills are essential to the engineers' employability level and are considered equally important as content knowledge certificates. Doyle lists a set of skills she considers to be determinants for the success of professionals. In this list, she includes skills such as communication, empathy, leadership, active listening, conflict management, negotiation, positive attitude, and teamwork [28]. The importance of these skills is also covered by the CDIO (Conceive Design Implement Operate) initiative, which is a framework that defines standards for engineering degrees [29]. These standards divide the skills to be developed by future engineers into three major categories: technical, knowledge, and reasoning; personal and professional skills and attributes; and interpersonal skills. Comparatively, Hernández-March and collaborators [26], in a study of the skills employers value in higher education graduates, organize these skills into four different groups: technical skills; interpersonal skills; cognitive skills; and methodological skills. The referred works were the starting point for defining the competences assessed by students' perceptions in this investigation.

In this study, perceptions were operationalized through students' self-assessment of a set of academic and interpersonal skills before and after confinement. This way, this work means to better understand how they feel affected by the years of lockdown.

2.3. Impact of COVID-19 on Students

In the last three years, many studies have focused on the impact of COVID-19 on young people and adolescents. Particularly with regard to students in higher education, research reports that the pandemic had an impact on several dimensions of students' lives, including their lifestyle, interpersonal competences, behaviors, emotions, feelings, and educational experiences [19,30–32]. Studies from different parts of the globe indicate some common issues experienced by students, such as a decrease in motivation due to social aspects and especially a lack of communication and interactions with teachers and peers [33–36]. In a systematic literature review carried out at the beginning of the pandemic crisis in 2020 on *The Impact of Social Isolation and Loneliness on the Mental Health of Children and Adolescents in the Context of COVID-19*, the authors found 83 articles that addressed this topic [37]. Sixty-three of these studies found a strong association between loneliness and mental health in children and adolescents, predicting that these problems may continue to arise for up to 9 years. One of the studies cited in the review states that children who had experienced forced isolation were five times more likely to need psychological support and to experience high levels of post-traumatic stress [38]. These results may indicate that it is expected that in the post-COVID-19 years, there will be a significant increase in mental health problems in young people. These results have been reinforced by a set of publications that, in recent years, have shown an increase in the prevalence of problems related to mental illness, such as stress, anxiety, lack of concentration, fear, sleep disorders, obesity, and depression [4,32,39–43].

Loneliness due to disease containment measures appears to be particularly problematic for young people, making them more vulnerable, namely because they feel deprived of the support of their peer group [37]. This may have more impact in educational settings where interpersonal relationships are more important [30].

In a study conducted by Ievers and collaborators on *The Impact of COVID-19 Restrictions upon Transversal Skills Development amongst Higher Education Students*, the authors found negative but also positive effects [44]. They mention positive impacts the developments that students report in their use of technology and digital literacy in general, which is corroborated by the findings of Gutierrez et al. [31]. In this way, with the right resources and support during the online classes, the authors argue students may even have the opportunity to improve their professional skills, such as communication, collaboration, self-efficacy, and digital skills. However, as these authors claim, these effects are likely to be due to students' exposure to online learning rather than to the lockdowns. Other positive impacts could be found in relation to citizenship, problem-solving skills, adaptability, self-reliance, and a small increase in inclination to listen to others and respect their point of view [19]. Still, other authors refer to COVID-19 as having a negative impact on students' problem-solving skills, time management ability, and teamwork, although their communication has been reinforced [19]. Ievers et al. [44] also mentioned a negative impact on the effective use of language, communication, the transmission of ideas, and confidence to engage in face-to-face communication right after the end of the pandemic restrictions. In fact, according to teachers' perception when returning to face-to-face classes, students had reinforced their communication not only with their peers but also with the faculty staff, including teachers (" ... more friendly in terms of social interaction that any other semester I have had ever.") [19]. Other authors also mention a positive aspect (referred by students) of the lockdowns, which was that they were able to have more flexibility by living, working, and studying in their cities and not having to commute anymore [33]. By changing some mindsets about how education must operate, society can emerge from this period with some positives to the learning process (using different and complementary ways of learning and practicing) and, on the other hand, help to reduce its ecological footprint in terms of transportation use, both contributing to sustainability [45].

3. Research Methodology

This study was conducted to better understand students' perception of some issues that might affect their performance not only in terms of learning the contents but also in developing social or technical competences. Its' main goal was to understand the cognitive and emotional effects COVID-19 had on students according to their own perceptions. The research question tackled in this work is: *"In which way the COVID-19 Pandemic with its lockdowns affected students' self-efficacy perception?"*.

To accomplish it, a survey was applied through an online questionnaire (google form) and a mixed analysis of the quantitative and qualitative data provided [46].

The dimensions in the study were based on the work of Alison Doyle [28] and Crawley [29]. We started by outlining two major dimensions that we defined as Interpersonal competences and academic competences. The first was subdivided into the categories: communication, empathy, focus, organization, creativity, adaptability, and emotions/attitudes. The second dimension considered the categories: work capacity, technical proficiency, theoretical knowledge, and management. For each category, several questions were constructed to address each item from different angles. Finally, an open question was added to allow students' own reflections and to gather richer data about students' main concerns. Each question assessed two moments: the students' perception of their self-assessment in relation to these competences before and after the pandemic.

The questionnaire was first developed by the authors of this work and then validated through two focus groups [46] from the higher education school of engineering where this study took place. The first group constituted nine researchers, and the second of eight students. Both group participants presented different backgrounds, representing varied contexts, and were involved in diverse engineering degrees from different levels (post-secondary technical degree, major degree, and master's degree). Each question was discussed in terms of its clarity, the terms used for the Likert scale of agreement were also discussed, and at the end, the overall questionnaire was addressed. The inputs and suggestions each group provided allowed the researchers to make some adjustments to the initial questionnaire to make the purpose of each question clearer and perfectly understandable. Some questions were redrawn or combined.

The anonymous questionnaire ended up with the following:

- Two initial questions to characterize the population (participants' present age and school year (before the pandemic and present, that is, in the 2022/23 curricular year).
- Thirty-eight closed questions with the option "not applicable" and a 5-level Likert agreement scale (1—minimum; 5—maximum); students had to answer each of these questions considering their perception before the pandemic (the curricular year 2019/20 before March) and their actual perception (after the pandemic).
- An open question to welcome students' comments or final remarks.

All the questions were mandatory, except the open question and the question about the school year (by students' suggestion, as some of them, before the pandemic, were not in a formal educational situation and felt awkward about it). The questionnaire (Appendix A) took about 15 min to be answered.

The survey was delivered during the first semester of 2022/23 in a school of engineering, covering the students' community, involving students from different curricular years, including post-secondary, major's, and master's degrees. Teachers acknowledged students of the importance and relevance of the study and that the data collected would only be used for the purposes of this research. Students' anonymity was assured, as well as the voluntary nature of their participation.

A Google questionnaire was used in the process of collecting the answers, which link most teachers shared with students during a class. Other teachers shared the link with their students via email or on their course MOODLE page. These data collections occurred between 8 December 2022 and 15 January 2023, corresponding to the last weeks of the semester. The collected data were treated with the Statistical Package for the Social Sciences (SPSS) software [47].

The identification of the factors was driven by a factor analysis (FA) procedure using SPSS. FA is a data reduction technique used to group a large number of (observed) variables into a smaller set of representative factors. So, our theoretical categories were analyzed in terms of their consistency in the questionnaire. Each of the analyzed factors complied with more than one category. From the factor analysis, a total of five factors were identified: **F1**—Communication and Empathy; **F2**—Focus and Personal Organization; **F3**—Teamwork and Individual Work Capacity; **F4**—Technical and Cognitive Resources Management; **F5**—Emotional Resources Management. Each one is analyzed in two temporal periods: before the pandemic and the present time, that is, after the pandemic.

The reliability of the questionnaire and factors of analysis assessed its internal consistency. Table 1 summarizes the five analyzed factors, including the questions incorporated in each one as well as the Cronbach alpha [48] for each factor (before (b) and after (a) the pandemic) for 483 valid answers.

Table 1. Students' questionnaire internal consistency analysis.

Factors	Questions (483 Valid Answers)	Cronbach Alpha	
		Before (B)	After (A)
F1—Communication and Empathy	1, 2, 3, 4, 5, 6, 7, 8, 9, 10	0.944	0.902
F2—Focus and Personal Organization	11, 12, 14, 15, 16, 17, 18, 19	0.943	0.922
F3—Teamwork and Individual Work Capacity	13, 20, 21, 22, 31, 32	0.937	0.888
F4—Technical and Cognitive Resources Management	23, 24, 25, 26, 27, 28, 29	0.929	0.891
F5—Emotional Resources Management	33, 34, 35, 36, 37	0.734	0.668

The former analysis shows internal consistency for all factors, although to a lesser extent for factor F5. To maintain this internal consistency at this high level, two of the 38 closed questions of the questionnaire were not used. In fact, these two questions were related to another category (creativity and adaptability) that theoretically did not fit into these five factors.

The qualitative analysis, related to the open question, was addressed using content analysis [46]. In general, students are not particularly attracted to this type of open question—it requires introspection, time, and effort—so when they answer, it is because they feel the need to express their feelings. The point was to identify the main ideas (besides different linguistic formulations) for each student and, when possible, relate them with the already represented factors. Each student's answer may express comments in more than one factor, so each answer may be spread out in several factors.

4. Data Analysis and Results

Of the 487 students who responded to the questionnaire, 483 were considered valid answers. Since the questionnaire was delivered to engineering students in a specific Engineering Institution, all respondents at this point (after the pandemic) are attending a college degree. Still, there were three students who answered "other" academic situation and one that answered "none", so these four answers were considered invalid.

4.1. Descriptive Analysis of the Collected Data

This valid sample constituted a diversified illustration in terms of age and in terms of academic background, and path, as illustrated in Figure 1.

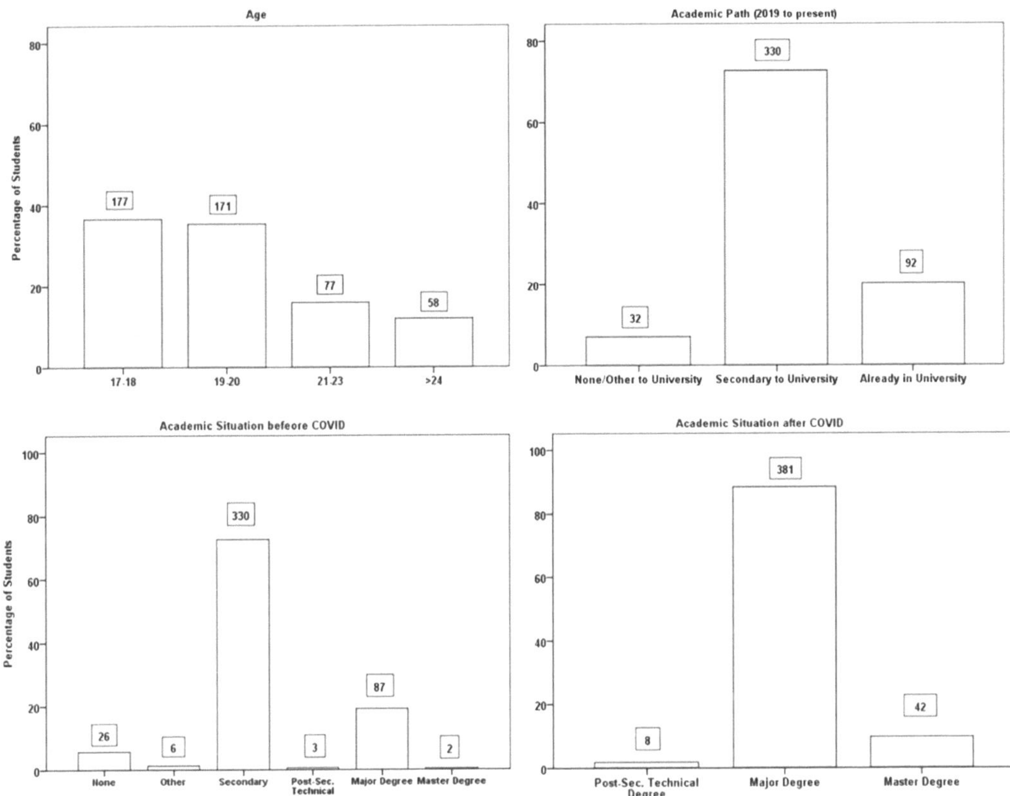

Figure 1. Sample characterization in terms of age distribution, academic path (**top**), and the academic situation before and after the pandemic (**bottom**).

In terms of age, the two larger groups had 17–18 years old (36.6%) and 19–20 years old (35.4%). The majority were students who were in high school at the time the pandemic broke out (68.3%). Some of them endured the pandemic for two years at high school, others for one year before entering university. Of those who endured the pandemic at university, 18.0% were in a major degree, 0.6% were in a post-secondary technical degree, and 0.4% attended a master's degree. Note that 5.4% were not studying at that time (that is, before the pandemic), and 1.2% were in "other" academic situations (which might mean nonacademic courses or non-degree courses). At the present time (after the pandemic), all students are attending a college degree. Most students were in a major degree (78.9%), some were already in a master's degree (8.7%), and others were in a post-secondary technical degree (1.7%).

The difference in the academic path (study level in 2019 until present) can influence students' perceptions since some moved from high school to university and may have experienced the pandemic period very differently from those who were already at university. So, to perceive these differences, the sample was split into three parts regarding their situation before the pandemic: the ones who did not attend school, those who were in high school, and those who were already at a university level.

Considering the 483 validated answers of each factor of analysis in both the pre- and post-pandemic phase (Table 2), the median value keeps unchanged in almost every factor, except for F2 (Focus and Personal Organization), where there is a downfall from 3.50 to 3.00.

Table 2. Median results in each factor of analysis for 483 valid answers (Likert scale 1–5).

Factors of Analysis		Median	Mean	Std. Deviation
F1—Communication and Empathy	Before	4.00	3.73	1.171
	After	4.00	3.78	0.979
F2—Focus and Personal Organization	Before	3.50	3.35	1.148
	After	3.00	3.14	1.075
F3—Teamwork and Individual Work Capacity	Before	4.00	3.98	1.138
	After	4.00	4.16	0.882
F4—Technical and Cognitive Resources Management	Before	4.00	3.50	1.140
	After	4.00	3.79	0.952
F5—Emotional Resources Management	Before	3.00	3.22	1.216
	After	3.00	2.89	1.160

Even though the analysis of the mean value of each factor is not so relevant, the corresponding calculation of the standard deviation indicates a higher coherence of students in answering the post-pandemic questions than the pre-pandemic ones. This is totally understandable since students were asked to think back almost three years to respond to the pre-pandemic state.

4.2. Direct Correlations between the Factors of Analysis

After addressing the test of normality (Kolmogorov-Smirnov) and establishing the non-normality of the data, Spearman correlations were addressed between variables. Some differences occurred between the pre- and post-pandemic period (before and after in Table 3). Considering all students' valid answers, all factors show significant moderate correlations between them, being the lowest ones with F5. Addressing the correlations between factors at the same period (before/before or after/after) that appear in Table 3 in shadow, almost all correlations decrease its intensity (still being moderate) after the pandemic, the larger differences between F2 with F3 and F3 with F5 (boxes with border in Table 3). This might mean students felt these factors more apart after the pandemic.

Table 3. Correlation between factors before and after COVID-19 pandemic (B—before; A—after).

Factors	Period	F1		F2		F3		F4		F5	
		B	A	B	A	B	A	B	A	B	A
F1	B		0.514 **	0.649 **	0.239 **	0.615 **	0.443 **	0.557 **	0.360 **	0.505 **	0.177 **
	A			0.299 **	0.638 **	0.446 **	0.600 **	0.369 **	0.582 **	0.313 **	0.497 **
F2	B				0.388 **	0.590 **	0.438 **	0.512 **	0.323 **	0.519 **	0.222 **
	A					0.344 **	0.508 **	0.291 **	0.556 **	0.208 **	0.512 **
F3	B						0.797 **	0.651 **	0.473 **	0.481 **	0.262 **
	A							0.510 **	0.610 **	0.332 **	0.358 **
F4	B								0.623 **	0.460 **	0.246 **
	A									0.317 **	0.405 **
F5	B										0.486 **
	A										

** Significant at the 0.01 level (2-tailed).

Analyzing the other correlations between factors before and after the pandemic stage (Table 3), other interesting correlations (from weak (<0.4) to strong (>0.7)) emerge. The first fact that stands out is the change in each factor between the two temporal moments (F1 before with F1 after, etc., indicated in bold in Table 3). Even though they are all significantly correlated, the highest ones are in F3 (strong correlation) and F4. This might mean that students felt less affected regarding Teamwork and Individual Work Capacity, and Technical and Cognitive Resources Management (in Table 2, it has already been shown that the median of these factors has not changed). Comparatively, the weakest one is found in F2 (weak correlation), which means that a greater percentage of students felt differently regarding their Focus and Personal Organization after the pandemic.

In relation to the correlations between factors at any temporal moment, a larger difference was found between F1 and F5, before and after (Table 3), which might mean that students percept these two factors (Communication and Empathy and Emotional Resources Management) further apart.

Considering the correlations between students' age and the factors, several very weak negative significant correlations appear before the pandemic (only F4 does not show this tendency) but no correlation after the pandemic. This could mean that the older the students were, the more confident they were in their capabilities. However, it also might mean that younger students really felt these factors stronger. Either way, these differences no longer appear after the pandemic. When analyzing the correlations between the factors and the study stage (secondary, post-secondary technical degree, major degree, master), no significant correlation appears, nor before nor after the pandemic.

4.3. Results Addressing Significant Differences Regarding Students Age and Study Path

After the descriptive analysis and the assessment of the correlation, the results were extracted to fully address the important issues under research. To perceive if there were statistically significant differences between the pre-pandemic stage (before) and the post-pandemic stage (after), a Wilcoxon nonparametric test was applied to the factors in the analysis.

First, an analysis was performed to understand if there were differences between the two moments in time (before and after). The results obtained for the 483 valid answers show (Table 4, first column) some significant differences: F2 and F5 decreased significantly, which means students felt negatively affected by the pandemic regarding their focus and personal organization and also in their emotional resources management. However, F3 and F4 show an increase. This means students percept a significant increase in relation to teamwork and individual work capacity and their technical and cognitive resources management. Only F1 does not show significant differences before and after the pandemic.

Table 4. Statistically significant differences in each factor for the total valid answers, by age interval and by study path.

Factors	All Students (483)	Age of Students (Years Old)				Study Path		
		17–18 (177)	19–20 (171)	21–23 (77)	>24 (58)	None/Other to University—*Largest Change* (32)	High School to University—*Large Change* (330)	Already in University—*Lesser Change* (92)
F1	**No difference**	No difference	**Decrease**	No difference	**Increase**	**Increase**	**Decrease**	No difference
F2	**Decrease**	Decrease	Decrease	Decrease	**Increase**	**Increase**	Decrease	Decrease
F3	**Increase**	No difference	No difference	Increase	Increase	Increase	No difference	No difference
F4	**Increase**	Increase	No difference	Increase	Increase	Increase	Increase	Increase
F5	**Decrease**	Decrease	Decrease	Decrease	**Increase**	**Increase**	Decrease	Decrease

However, these differences might be felt differently by distinctive groups of students, so the analysis of the significant differences was addressed by age and by their study path. This is also shown in Table 4, where the modifications (in comparison with the group of all students) are highlighted in shadow. For this analysis, the students were grouped by age intervals or by their study path. For 29 students, the information about the study path was not available (in the analysis, was considered as missing data). So, the major results per factor are:

- F1: This factor is where the major differences occur in relation to the whole group of students. The group of 19–20 years old and the group who suffered the *largest change* in their study path (from high school to university) show a statistically significant change, considering their capacities in relation to communication and empathy decreased. On the other hand, older students have a different perspective, showing an improvement, which is also found in the group of students who suffered the *largest change* of all (the ones who were not studying or attended another type of education before the pandemic).
- F2: Only the older students and the ones who suffered the *large change* in their study path have a different perspective, considering their capacities in relation to concentration and personal organization improved.
- F3: Again, only older students (more than 21 years old) and the ones who suffered the *largest change* in the study path show the increase observed in the whole group, considering that their capacities in relation to teamwork and work capacity have improved, even though in this group the observed difference is more meaningful than in the whole group.
- F4: This factor is almost the same in every group, showing a significant increase in precepted capabilities regarding technical and cognitive resources management. Only in the group between 19–20, no significant difference between before and after the pandemic was found.
- F5: Again, all groups show the tendency to decrease their perception regarding their emotional resources management. Only the older group of students and the ones who suffered the *largest change* in their study path expressed an improvement after the pandemic.

It becomes clear that the older students and the group who suffered the *largest change* have similar perceptions. The overlap of these two groups represents 41% in relation to the age group (24 of these older students were part of the group that undertook the *largest change* in their study path).

The fact that these two groups increased their perception in all factors might indicate a difference in adulthood, more life experience, and, in the case of students who were not attending school, the increase of their capacities in factors related to the dimension of academic competences is understandable.

Students who have moved to a university level consider that they have only increased their perception of their capabilities in relation to technical and cognitive resources management (F4) and maintained the result of F3 (teamwork). This result might not be directly related to the pandemic but be a natural course of events in their lives due to the differences between the two worlds (secondary and university levels). It could also be related to students' lower level of maturity. Students who were already at university showed similar behavior apart from the fact that they did not show a significant change in F1 (communication and empathy), probably because they already knew their colleagues from college, and it was easier for them than for the previous group.

In terms of age, the group that shows a larger decrease in these factors' perceptions is the group between 19–20 years old (this group of students was in the majority the ones who entered university in the second year of the pandemic). Somehow this was not what was primarily expected. The younger ones (17–18 years old) were the students who endured the totality of the pandemic in high school. However, this result may indicate that students who endured the pandemic only in high school only experienced one kind of change. Students who experienced pandemic restrictions in both high school and university had to adapt to two ways of working during the pandemic. The results of the qualitative analysis obtained from the content analysis of the open question show support for some of the previous results, namely with factor F5. The overall analysis of the 30 answers (6.2% of the respondents) allowed us to identify some issues directly related to the questionnaire's main goal. Table 5 shows these results summarized by groups, split according to their impact (negative, positive, neutral).

Table 5. Content analysis of the open question.

Impact	Issues	Number of Students	Group Identification	
			Age (Interval)	Study Path
Negative	Socialization difficulties (F5)	3	2 in (19, 20), 1 in >24	1 missing; 1 none/other to university; 1 from high school to university
	Depression/anxiety (F5)	4	2 in (19, 20), 2 in (21–23), 1 in >24	2 from high school to university, 1 already in university, 1 from major to master's degree
	Difficulty on focusing and self-discipline (F2)	4	3 in (17, 18), 1 in >24	from high school to university
	Anger and lack of patience (F1, F5)	1	(17, 18)	from high school to university
	Lack of confidence and motivation (F5)	2	1 in (17, 18), 1 in >24	1 missing; 1 from high school to university
Positive	Personal organization (F2)	1	>24	1 already in university
	Resources (F4)	2	1 in (17, 18), 1 in >24	1 missing; 1 from high school to university
Neutral	No changes	5	2 in (17, 18), 1 in (19, 20), 2 in >24	2 missing; 3 from high school to university
	Difficulty in differentiate from other changes	5	4 in (19, 20), 1 in (21–23),	4 from high school to university 1 from major to master's degree
	Other comments	2	1 in (19, 20), 1 in (21–23),	1 missing; 1 from high school to university

The most referred aspect was *"negative emotions"* which accounted for 11 answers; students expressed different emotions: "I felt compelled to interrupt my academic year due to depression/anxiety", " ... had to take meds to control it", "nowadays I cannot focus for more than some minutes ... ", "the major pandemic effect on me was the difficulties in socialization with colleagues ... ". Considering the aspects that had a positive impact, two students generally refer to the pandemic helping them to learn in several ways or to use their own words, "they have learned from it". Other students are more precise and clearly identify the aspects the pandemic helped them to improve, as detailed in Table 5: "resources for remote work", "I can manage my time much better now ... ".

Some students expressed they feel good and/or do not feel the pandemic has affected their behavior (*"no changes"*); some expressed it was difficult for them to understand the impact the pandemic had on their behavior as they have experienced a change in their study path and that modification also had an enormous impact in their lives. In *"other comments"*, it was considered the feeling of gratitude two students have expressed for having the chance to be heard.

5. Discussion

Students perceived the pandemic restrictions differently according to their age and specific parameters. In relation to F1—Communication and Empathy, some authors point out that the young generation is used to communicating regularly online, even though the lack of face-to-face contact between peers could affect empathy. In this research, in relation to this factor, no significant differences were found in the student's assessment before and after the lockdowns, seeming to indicate that the isolation to which they were subjected did not impact on communication and empathy skills. In a way, these results go against those obtained by Ievers et al. [44], according to whom (and also according to students' perspective), the isolation caused by successive lockdowns provoked a small increase in inclination to listen to others and respect their point of views [44]. It also contradicts the work of Becker et al. [19], in which teachers state students' communication ability clearly increased, being the students more friendly between themselves and with others.

Regarding factor F2—Focus and Personal Organization, the results show a significant decrease, indicating that students feel that they have lost skills related to the ability to focus on tasks, time management, and work organization. Indeed, many studies report consequences at the emotional level and on students' ability to concentrate, including time management issues [4,39–43]. Emotional disorders such as anxiety and stress may affect the subjects' ability to concentrate on tasks and, consequently, their focus and personal organization. Still, in relation to this factor, the older students and the ones who were not studying before the pandemic considered that their capacities in relation to focus and personal organization improved. This finding may be due to the fact that these students feel that they would be willing to make greater efforts to succeed in the face of such a significant change in their lives; or that these students, being more experienced and mature, may develop more positive self-efficacy beliefs. In contrast to the previous factors, students reported an increase in their competencies related to F3 and F4—Teamwork and Individual Work Capacity and Technical and Cognitive Resources Management, respectively. Our study showed a more significant increase in F4 than in F3, being the work capacity increase more pronounced in those who were not attending school during the pandemic or in the group of older students. These results are in accordance with other works that reported an enhancement in collaboration (teamwork), digital skills, and in self-efficacy [31]. Other studies also claim that the online environment favored some students who were more self-motivated or with higher self-regulating capacities [33,49].

In relation to F5—Emotional Resources Management, the literature supports that students' mental health has been severely affected, being one of the most cited factors. The reported social impact during the pandemic [32–34,36] might have left repercussions that students still need to overcome, such as anxiety, depression, sleep disorder or poor sleep. The findings of these works corroborate this study's results that there was a significant decrease in their emotional resources management.

The increase found in the group of students who were not in school during the pandemic may be consistent with Alexa et al. [33], who suggest students may have felt more flexibility in balancing their work and school lives. This might have been a reason for some of them to go back to school.

6. Conclusions

A better understanding of how students are currently coping with their learning, namely identifying possible gaps in their learning during COVID-19, deficits in the development of competences, or, on the contrary, the development of other skills or auxiliary tools, can indeed provide valuable information to teachers, who can thus better tailor their teaching to the student's needs. This more targeted teaching with a view to improving academic success can, in the long run, lead to more sustainable teaching.

With this work, some important aspects were more clearly identified, namely some positive aspects that teachers might consider continuing to use, such as auxiliary/complementary tools that allow students to practice more autonomously. However, these resources should be made available with an organizational plan and explanations of how and when to use them. This derives from the results pointing that globally, with the pandemic and confinement period, students percept they improved their teamwork and individual work capacity and their technical and cognitive resources management. However, their ability to focus and personal organization, as well as their emotional resources management, have deteriorated.

Students who were already at university show a smaller impact than those who have gone from high school to university.

Students who did not attend school before the pandemic and the group of older students (over 24 years old) show an improvement in their perception of their abilities in all factors, which means that this group is better able to cope with those restrictions, and some of them expressed they have learned from it. Students who were in high school before the pandemic underwent a greater change in their educational environment and probably other aspects of their life (when entering university, these students usually undergo changes in maturity, from being more independent, sometimes leaving home, changing city of residence, etc.), which might have influenced the change of perception observed in this group. This was also mentioned by some students, some of them clearly indicating this fact may have blurred their perception of the real impact of the pandemic. However, the only difference with the group of students who already were in university is strictly in relation to communication and empathy, where they show a decrease. Interesting evidence emerged differentiating students who had experienced the two years of the pandemic in high school from those who had experienced one year of restrictions and online classes in each level of education. The group who moved to the university showed more difficulty with communication and empathy issues, which may be due to the two major changes and adaptations they had to endure, one adapting to the online teaching and resources in high school and the other in university.

So, answering our research question, *"In which way the COVID-19 Pandemic with its lockdowns affected students' self-efficacy perception?"* this work shows that older students were less affected than younger ones, indicating a greater ability to adapt and cope with the pandemic restrains. In general, students seem to have been able to be more autonomous as they were capable of working and developing their cognitive resources, but their emotional state and ability to focus were reduced. All these factors influenced students' self-efficacy perception. Although students who changed their education level were the ones who suffered more, there was a significant difference between those who experienced these changes in teaching only at one level of education and those who experienced them at both levels (high school and university), with the former being the most negatively affected.

This study has some limitations, namely the fact that students were asked to answer the questions about their pre- and post-pandemic competencies in a single time point, with the presumed recall bias that this might introduce. Furthermore, the fact that the students are only from one educational institution, as well as the size of the sample, reduces the possibility of extrapolating these results. The fact that this study is based only on self-report measures is also one of the limitations of this study; moreover, given the age of some participants, it is impossible to categorically affirm if the evolution recognized by the results of this study is derived from the COVID-19 restrictions or the natural increase in maturity of those students.

To strengthen the consistency of these results, it would be interesting to carry out a longitudinal study that would allow monitoring of the evolution of these students' perceptions and possibly corroborate the results with an analysis of the teachers' perceptions of their students' learning.

Author Contributions: Conceptualization, C.V. and A.R.C.; data curation, C.V. and N.L.; formal analysis, N.L. and A.R.C.; methodology, N.L.; supervision, C.V.; writing—original draft, C.V., N.L. and A.R.C.; writing—review and editing, C.V., N.L. and A.R.C. All authors have read and agreed to the published version of the manuscript.

Funding: The authors would like to acknowledge the support provided by the Portuguese Foundation for Science and Technology Project, FCT UIDB/04730/2020.

Institutional Review Board Statement: Ethical review and approval were waived for this study due to ISEP/P.PORTO not having a formal Ethics Committee. To ensure ethical concerns, the survey was volunteer and anonymous. In the first section of the questionnaire, all participants were fully informed about the guarantee of anonymity, the research objectives, how the data would be used only for research purposes and the authors' contact details.

Acknowledgments: The authors would like to thank the students who completed the questionnaire and shared their insights. A special appreciation to the students and colleagues who participated in each focus group session and all the colleagues who delivered the questionnaire to their students.

Conflicts of Interest: The authors declare no conflict of interest.

Appendix A. Questionnaire on Students' Perceptions Before and After the Pandemic

This study aims to better understand possible changes felt by students after the pandemic period and the confinements they were subjected to. The idea is to compare their perception before the pandemic/confinements with the current situation (present time).

The data collected are anonymous and will only be used for scientific research. The results of this survey may be sent to interested parties by email to the authors.

The questionnaire has 38 questions and can be answered in about 15 min.

The scale of answers varies from 1 to 5, where 1 corresponds to completely false (not at all) and 5 completely true (yes, completely)

Thank you for your valuable contribution!

Appendix A.1. Identification of the Profile

Age:______

Curricular year you attended/attend (before the pandemic/at present):

	In 2019/2020 (1ˢᵗ confinement was in March 2020)	At Present Time
10°ano	☐	☐
11°ano	☐	☐
12°ano	☐	☐
Ano Zero	☐	☐
1°ano CTESP	☐	☐
2°ano CTESP	☐	☐
1°ano licenciatura	☐	☐
2°ano licenciatura	☐	☐
3°ano licenciatura	☐	☐
1°ano mestrado	☐	☐
2°ano mestrado	☐	☐
outro	☐	☐
nenhum	☐	☐

Appendix A.2. Questionnaire

(an example of the answer format is presented, equal for the 38 questions)

Before the Lockdowns:	not applicable	1	2	3	4	5
At Present Time:	not applicable	1	2	3	4	5

*where **1** corresponds to completely false (**not at all**) and **5** completely true (**yes, completely**)*

1. I am able to participate actively in class
2. I am able to do work presentations
3. I am able to think and try to answer teachers' questions
4. I am able to communicate with colleagues in teamwork
5. I am able to defend my point of view in class
6. I am able to understand and/or argue with colleagues' points of view

7. I have the patience to be in a whole class and deal with classmates and teachers
8. I feel able to support colleagues with more difficulties
9. I get along easily with colleagues
10. I feel comfortable talking to teachers
11. I am able to manage my time
12. I feel motivated to work in class
13. I am able to make teamwork decisions
14. I am able to keep away from social networks during work periods
15. I am able to concentrate on the tasks I am doing
16. I am able to stay focused during a complete task
17. I am able to keep my attention when I am listening to others
18. I am able to keep focused on my studies
19. I manage not to be distracted in class
20. I can be open-minded to ideas different from my own
21. I can respect hierarchy and teamwork distribution of tasks
22. I can respect the ideas of my colleagues
23. I can easily write a text work
24. I can read, interpret, and understand an academic text
25. I can find ICT (Information and Communication Technology) tools that help me in my work
26. I can use and master the ICT tools I need
27. I can master varied resources to work on experimental concepts (e.g., simulators, remote laboratories …)
28. I can easily adapt to new situations
29. I can critically analyze information and/or the results of an assignment
30. I can be flexible in the way I work
31. I am able to create team spirit in teamwork
32. I can easily collaborate with my teammates
33. I am able to manage anxiety
34. I have little motivation to leave home for school
35. I am able to manage stress
36. I can balance my studies with my social life
37. In my daily life I often feel pessimistic and sad
38. I can be creative in my daily life

References

1. Auxier, B.; Andresen, M. As Schools Close due to the Coronavirus, Some U.S. Students Face a Digital 'Homework Gap'. Pew Research Center. 2020. Available online: https://www.pewresearch.org/fact-tank/2020/03/16/as-schools-close-due-to-the-coronavirus-some-u-s-students-face-a-digital-homework-gap/ (accessed on 17 March 2023).
2. Coleman, T. *Digital Divide in UK Education during COVID-19 Pandemic: Literature Review*; Cambridge Assessment: Cambridge, UK, 2021.
3. Curelaru, M.; Curelaru, V.; Cristea, M. Students' Perceptions of Online Learning during COVID-19 Pandemic: A Qualitative Approach. *Sustainability* **2022**, *14*, 8138. [CrossRef]
4. Kim, N.H.; Lee, J.M.; Yoo, E. How the COVID-19 Pandemic Has Changed Adolescent Health: Physical Activity, Sleep, Obesity, and Mental Health. *Int. J. Environ. Res. Public Health* **2022**, *19*, 9224. [CrossRef] [PubMed]
5. Mannion, R.; Konteh, F.H.; Jacobs, R. Impact of COVID-19 in mental health trusts. *J. Health Serv. Res. Policy* **2022**, *28*, 119–127. [CrossRef]
6. UNESCO. Education in a Post Covid World: Nine Ideas for Public Action International Commission on the Futures of Education. 2020. Available online: https://en.unesco.org/news/education-post-covid-world-nine-ideas-public-action (accessed on 23 October 2022).
7. Benito, Á.; Yenisey, K.D.; Khanna, K.; Masis, M.F.; Monge, R.M.; Tugtan, M.A.; Araya, L.D.V.; Vig, R. Changes that should remain in higher education post COVID-19: A mixed-methods analysis of the experiences at three universities. *High. Learn. Res. Commun.* **2021**, *11*, 51–75. [CrossRef]
8. García-Peñalvo, F.J.; Corell, A.; Abella-García, V.; Grande, M. La evaluación online en la educación superior en tiempos de la COVID-19. *Educ. Knowl. Soc. EKS* **2020**, *21*, 26. [CrossRef]
9. Nelson, B. The positive effects of COVID-19. *BMJ* **2020**, *369*, m1785. [CrossRef] [PubMed]

10. Viegas, C.; Lima, N. Promoting Students' Learning and Involvement Under Demanding Remote Environments. In *Learning with Technologies and Technologies in Learning. Lecture Notes in Networks and Systems*; Auer, M.E., Pester, A., May, D., Eds.; Springer International Publishing: Berlin/Heidelberg, Germany, 2022; Volume 456, pp. 603–628. [CrossRef]

11. Asif, M.; Khan, M.A.; Habib, S. Students' Perception towards New Face of Education during This Unprecedented Phase of COVID-19 Outbreak: An Empirical Study of Higher Educational Institutions in Saudi Arabia. *Eur. J. Investig. Health Psychol. Educ.* **2022**, *12*, 835–853. [CrossRef] [PubMed]

12. García-Peñalvo, F.J. El Sistema Universitario ante la COVID-19: Corto, Medio y Largo Plazo. Universídad. 2020. Available online: https://bit.ly/2YPUeXU (accessed on 28 March 2023).

13. Zakaria, S.; Samat, M.F.; Abdullah, A. The Role of Relational Trust during COVID-19 Pandemic Among University Students. *Int. J. Bus. Soc.* **2022**, *23*, 1286–1296. [CrossRef]

14. Cleofas, J.V.; Albao, B.T.; Dayrit, J.C.S. Emerging Adulthood Uses and Gratifications of Social Media During the COVID-19 Pandemic: A Mixed Methods Study Among Filipino College Students. *Emerg. Adulthood* **2022**, *10*, 1602–1616. [CrossRef]

15. Hess, F.M. Education after the Pandemic. National Affairs. 2022. Available online: https://www.realclearpolicy.com/2022/01/05/education_after_the_pandemic_810546.html (accessed on 28 March 2023).

16. Muthuprasad, T.; Aiswarya, S.; Aditya, K.S.; Jha, G.K. Students' perception and preference for online education in India during COVID-19 pandemic. *Soc. Sci. Humanit. Open* **2021**, *3*, 100101. [CrossRef]

17. Silva, O.; Sousa, Á. Perception of Teachers and Students about Teaching and Learning in the Period of COVID-19 Pandemic. In Proceedings of the 13th Annual International Conference of Education, Research and Innovation, Online, 9–10 November 2020; pp. 4832–4838. [CrossRef]

18. Stoian, C.E.; Fărcașiu, M.A.; Dragomir, G.-M.; Gherheș, V. Transition from Online to Face-to-Face Education after COVID-19: The Benefits of Online Education from Students' Perspective. *Sustainability* **2022**, *14*, 12812. [CrossRef]

19. Becker, T.B.; Fenton, J.I.; Nikolai, M.; Comstock, S.S.; Swada, J.G.; Weatherspoon, L.J.; Tucker, R.M. The impact of COVID-19 on student learning during the transition from remote to in-person learning: Using mind mapping to identify and address faculty concerns. *Adv. Physiol. Educ.* **2022**, *46*, 742–751. [CrossRef]

20. Bandura, A.; Locke, E.A. Negative self-efficacy and goal effects revisited. *J. Appl. Psychol.* **2003**, *88*, 87–99. [CrossRef]

21. Costa, A.R.; Araújo, A.M.; Almeida, L.S. Relação entre a percepção da autoeficácia acadèmica e o engagement de estudantes de engenharia. *Int. J. Dev. Educ. Psychol. Rev. INFAD Psicol.* **2016**, *2*, 307. [CrossRef]

22. Hunsu, N.J.; Olaogun, O.P.; Oje, A.V.; Carnell, P.H.; Morkos, B. Investigating students' motivational goals and self-efficacy and task beliefs in relationship to course attendance and prior knowledge in an undergraduate statics course. *J. Eng. Educ.* **2023**, *112*, 108–124. [CrossRef]

23. Salanova, M.; Martínez, I.; Llorens, S. Success breeds success, especially when self-efficacy is related with an internal attribution of causality. *Estud. Psicol.* **2012**, *33*, 151–165. [CrossRef]

24. Zimmerman, B.J. Self-Efficacy: An Essential Motive to Learn. *Contemp. Educ. Psychol.* **2000**, *25*, 82–91. [CrossRef] [PubMed]

25. Froyd, J.E.; Wankat, P.C.; Smith, K.A. Five major shifts in 100 years of engineering education. *Proc. IEEE* **2012**, *100*, 1344–1360. [CrossRef]

26. Hernandez-March, J.; Martin Del Peso, M.; Leguey, S. Graduates' skills and higher education: The employers' perspective. *Tert. Educ. Manag.* **2009**, *15*, 1–16. [CrossRef]

27. Stiwne, E.E.; Jungert, T. Engineering students' experiences of transition from study to work. *J. Educ. Work.* **2010**, *23*, 417–437. [CrossRef]

28. Doyle, A. Top Interpersonal Skills That Employers Value. The Balance. 5 May 2022. Available online: https://www.thebalancemoney.com/interpersonal-skills-list-2063724 (accessed on 12 October 2022).

29. Crawley, E.F. The CDIO Syllabus A Statement of Goals for Undergraduate Engineering Education. 2001. Available online: http://www.cdio.org (accessed on 12 October 2022).

30. Gadi, N.; Saleh, S.; Johnson, J.-A.; Trinidade, A. The impact of the COVID-19 pandemic on the lifestyle and behaviours, mental health and education of students studying healthcare-related courses at a British university. *BMC Med. Educ.* **2022**, *22*, 115. [CrossRef] [PubMed]

31. Gutierrez, K.S.; Kidd, J.J.; Lee, M.J.; Pazos, P.; Kaipa, K.; Ringleb, S.I.; Ayala, O. Undergraduate Engineering and Education Students Reflect on Their Interdisciplinary Teamwork Experiences Following Transition to Virtual Instruction Caused by COVID-19. *Educ. Sci.* **2022**, *12*, 623. [CrossRef]

32. Puteikis, K.; Mameniškytė, A.; Mameniškienė, R. Sleep Quality, Mental Health and Learning among High School Students after Reopening Schools during the COVID-19 Pandemic: Results of a Cross-Sectional Online Survey. *Int. J. Environ. Res. Public Health* **2022**, *19*, 2553. [CrossRef] [PubMed]

33. Alexa, L.; Pîslaru, M.; Avasilcăi, S.; Lucescu, L.; Bujor, A.; Avram, E. Exploring Romanian Engineering Students' Perceptions of COVID-19 Emergency e-Learning Situation. A Mixed-Method Case Study. *Electron. J. E-Learn.* **2022**, *20*, 19–35. [CrossRef]

34. Alghamdi, A.A. Impact of the COVID-19 pandemic on the social and educational aspects of Saudi university students' lives. *PLoS ONE* **2021**, *16*, e0250026. [CrossRef] [PubMed]

35. Morán-Soto, G.; Marsh, A.; González Peña, O.I.; Sheppard, M.; Gómez-Quiñones, J.I.; Benson, L.C. Effect of the COVID-19 Pandemic on the Sense of Belonging in Higher Education for STEM Students in the United States and Mexico. *Sustainability* **2022**, *14*, 6627. [CrossRef]

36. Tomić, I.; Pinćjer, I.; Dedijer, S.; Adamović, S. Online learning during COVID-19 pandemic as perceived by the students of Graphic engineering and design. *J. Graph. Eng. Des.* **2022**, *13*, 15–20. [CrossRef]
37. Loades, M.E.; Chatburn, E.; Higson-Sweeney, N.; Reynolds, S.; Shafran, R.; Bridgen, A.; Linney, C.; McManus, M.N.; Borwick, C.; Crawley, E. Rapid Systematic Review: The Impact of Social Isolation and Loneliness on the Mental Health of Children and Adolescents in the Context of COVID-19 | Elsevier Enhanced Reader. *J. Am. Acad. Child Adolesc. Psychiatry* **2020**, *59*, 1218–1239. [CrossRef]
38. Sprang, G.; Silman, M. Posttraumatic stress disorder in parents and youth after health-related disasters. *Disaster Med. Public Health Prep.* **2013**, *7*, 105–110. [CrossRef]
39. Dadaczynski, K.; Okan, O.; Messer, M.; Rathmann, K. University students' sense of coherence, future worries and mental health: Findings from the German COVID-HL-survey. *Health Promot. Int.* **2022**, *37*, daab070. [CrossRef] [PubMed]
40. Idoiaga, N.; Legorburu, I.; Ozamiz-Etxebarria, N.; Lipnicki, D.M.; Villagrasa, B.; Santabárbara, J. Prevalence of Post-Traumatic Stress Disorder (PTSD) in University Students during the COVID-19 Pandemic: A Meta-Analysis Attending SDG 3 and 4 of the 2030 Agenda. *Sustainability* **2022**, *14*, 7914. [CrossRef]
41. Mohammed, M.A.; Memmedova, K. Prevalence of Mental Health Problems among Iraqi University Students during the COVID-19 Pandemic. *Sustainability* **2023**, *15*, 1746. [CrossRef]
42. Schwartz, K.D.; Exner-Cortens, D.; McMorris, C.A.; Makarenko, E.; Arnold, P.; Van Bavel, M.; Williams, S.; Canfield, R. COVID-19 and Student Well-Being: Stress and Mental Health during Return-to-School. *Can. J. Sch. Psychol.* **2021**, *36*, 166–185. Available online: https://search.ebscohost.com/login.aspx?direct=true&db=eric&AN=EJ1295492&site=eds-live (accessed on 12 October 2022). [CrossRef] [PubMed]
43. Zhang, Y.; Guo, L.; Zuo, H. Short report: COVID-19 related knowledge, anxiety, and attitude towards the back-to-school arrangement among college students in China: A cross-sectional study. *Psychol. Health Med.* **2022**, 1–7. [CrossRef] [PubMed]
44. Ievers, M.; Cummins, B.; Ballentine, M. The Impact of COVID-19 Restrictions upon Transversal Skills Development amongst Higher Education Students. *Teach. Educ. Adv. Netw. J.* **2022**, *14*, 95–110.
45. Viegas, C.; Lima, N.; Felgueiras, M.C.; Marques, M.A.; Alves, G.R.; Costa, R.; Fidalgo, A. How may teaching contribute to sustainability in a small scale but with wide use? *TEEM* **2021**, *21*, 794–799.
46. Cohen, L.; Manion, L.; Morrison, K. *Research Methods in Education*, 6th ed.; Routledge Falmer: New York, NY, USA, 2007.
47. Marôco, J. *Análise Estatística com o SPSS Statistics*, 7th ed.; ReportNumber, Lda: Pêro Pinheiro, Portugal, 2018.
48. Pestana, M.H.; Gajeiro, J.N. *Análise de Dados para Ciências Sociais A Complementaridade do SPSS*, 6th ed.; Edícões Sílabo: Lisbon, Portugal, 2014.
49. Ahag, P.; Hsu, Y.J.; Olsson, L.; Sundberg, L. The Impact of SARS-CoV-2 on Engineering Education: Student Perceptions from Three Countries. In Proceedings of the 2020 IEEE International Conference on Industrial Engineering and Engineering Management (IEEM), Singapore, 14–17 December 2020; pp. 1266–1270. [CrossRef]

Article

Empirical Analysis of University–Industry Collaboration in Postgraduate Education: A Case Study of Chinese Universities of Applied Sciences

Ye Zhang [1] and Xinrong Chen [2,*]

[1] School of International Exchange, Shanghai Polytechnic University, Shanghai 201209, China
[2] Academy for Engineering and Technology, Fudan University, Shanghai 200433, China
* Correspondence: chenxinrong@fudan.edu.cn

Abstract: The training of professional degree postgraduates in universities of applied sciences is essential in meeting the needs of industry and society. However, there are challenges, such as structural unemployment and poor quality of application-oriented higher education, which can be addressed through university–industry collaboration. This study investigates the perceptions of professional degree postgraduates towards university–industry collaboration and identifies the areas of dissatisfaction. The findings show that postgraduates have a high degree of recognition of university–industry collaboration, but the main dissatisfaction lies in the alignment between enterprise practice and professional learning. To enhance the quality of training, universities should prioritize practice-oriented approaches that emphasize engineering practice throughout the entire training process, optimize the university–industry collaboration mechanism, and strengthen the construction of "double supervisor" faculties. These strategies can comprehensively enhance the training quality of professional degree postgraduates in universities of applied sciences, and ultimately improve their employability and contribution to society.

Keywords: postgraduate education; university–industry collaboration; universities of applied sciences; China

check for updates

Citation: Zhang, Y.; Chen, X. Empirical Analysis of University–Industry Collaboration in Postgraduate Education: A Case Study of Chinese Universities of Applied Sciences. *Sustainability* **2023**, *15*, 6252. https://doi.org/10.3390/su15076252

Academic Editors: Clara Viegas and Natércia Lima

Received: 19 February 2023
Revised: 17 March 2023
Accepted: 4 April 2023
Published: 5 April 2023

1. Introduction

Since the Academic Degrees Committee of the State Council officially introduced the term "professional degree" in 1996, China's postgraduate education has formed a postgraduate education system in which "academic" and "professional" degrees coexist. A "professional degree" is a type of degree set up to meet the needs of specific vocational fields in society and educate high-level application-oriented professionals with clear vocational and application orientations and strong professional abilities and qualities capable of engaging in practical work creatively [1]. The educational goal of professional degree graduate students is to educate application-oriented professionals who have certain theoretical research skills and abilities and can adapt to the needs of specific industries and fields [2].

In September 2020, the Academic Degrees Committee of the State Council and the Ministry of Education jointly released the Program for the Development of Professional Degree Graduate Education (2020–2025), marking a new stage of development of professional degree graduate education in China. It is stated in the program that professional degree graduate education mainly focuses on the needs of specific vocational fields in society and educates high-level application-oriented professionals with strong professional abilities and professionalism who can creatively engage in practical work.

At a time when the division of labor in society is becoming increasingly refined and specialized, professional degree graduate education has the unique advantage of meeting the diversified needs of professionals [3]. In November 2021, the Ministry of Education in China issued the Notice of the Academic Degrees Committee of the State Council on the

List of Additional Doctoral and Master's Degree Authorization Points to be Audited in 2020. In the results of the degree authorization audit, more than 1500 new master's degree authorization points were added, of which, 1115 were professional degrees, accounting for more than 70%. In 2021, the number of enrollments of professional degree master's students accounted for 60.8% of the total enrollment of master's students. Among them, the number of enrollments of professional degree master's students in engineering ranked first, becoming the largest and widest type of professional degree education in China.

At present, some universities of applied sciences in China adopt university–industry collaboration for postgraduate education. Students raise problems in enterprise practice, discover solutions in practice and theoretical study, and finally test research results in practice, which is an exploration and attempt to educate professional degree graduate students in universities of applied sciences [4,5]. This mode takes engineering students through the whole process of postgraduate education and emphasizes the full coverage of engineering practice and uninterrupted engineering training [6]. In their first year, students receive centralized theoretical study at university, and then go to enterprises directly to get familiar with relevant technologies and complete a 10–12 month internship under the guidance of enterprise supervisors. In their second year, students return to university to continue studying theoretical knowledge and carry out research under the guidance of university supervisors in combination with the projects carried out during the internship in enterprises. In their third year, students put research results into engineering practice. Students find and ask questions in real environments of engineering practice sites, determine research plans, form research topics, and develop feasible solutions or designs for solving problems under the dual guidance of enterprise supervisors and university supervisors [7].

In this paper, Section 2 provides a comprehensive literature review followed by Section 3, which outlines the methodology adopted for our study. In Section 4, we present the results of our analysis, including the characteristics of the sample, descriptive statistics, and linear regression analysis. Finally, we discuss the findings and draw conclusions based on the results obtained.

2. Literature Review

2.1. University–Business Cooperation

University–business cooperation, also known as industry–academia collaboration, is an area of research that has gained increasing attention in recent years [8]. The main objective of such cooperation is to facilitate knowledge transfer between universities and businesses, leading to the development of innovative products and services [9,10].

Research in this area has identified a number of key factors that can facilitate or hinder successful university–business cooperation [11,12]. For example, trust between the two parties, effective communication, and a shared understanding of goals are seen as important factors that can contribute to successful collaborations [13]. On the other hand, differences in organizational culture, conflicting priorities, and intellectual property issues are identified as potential barriers to collaboration [14].

Studies have also examined the different forms that university–business cooperation can take, including joint research projects, licensing agreements, and industry-sponsored research [15,16]. The effectiveness of these different forms of collaboration varies depending on factors such as the type of industry, the nature of the research, and the goals of the collaboration [16].

The current research suggests that university–business cooperation can be a valuable means of promoting innovation and economic development, particularly in the science, technology, engineering, and mathematics (STEM) fields [17]. However, it also highlights the importance of effective communication, trust, and a shared vision in order to realize the full potential of such collaborations [18].

In recent years, there has been a growing recognition of the benefits of university–business cooperation for both academic and business communities [19,20]. For universities, such collaborations can provide access to new research opportunities, funding, and industry expertise, as well as opportunities to apply research in practical settings. For businesses, university partnerships can provide access to cutting-edge research and talent as well as the opportunity to collaborate with leading experts in their field.

As a result, there has been an increasing emphasis on fostering university–business collaboration, with initiatives such as industry–academic consortia, joint research centers, and technology transfer offices [20]. These initiatives aim to facilitate partnerships between universities and businesses, providing a framework for collaboration and ensuring that research outcomes are relevant and applicable to industry needs.

However, while the potential benefits of university–business cooperation are clear, there are also challenges to be overcome. For example, there can be cultural differences between academic and business environments, as well as differences in the pace of decision-making and approaches to risk. In addition, the ownership of intellectual property and the sharing of benefits from collaborative projects can be complex issues that require careful negotiation [21]. To address these challenges, there is a need for effective communication and clear understanding of the objectives and expectations of all parties involved. This includes establishing effective governance structures, identifying shared goals, and ensuring that the benefits of collaboration are fairly distributed.

In conclusion, university–business cooperation is an area of research that is rapidly evolving and has the potential to drive innovation and economic growth. While there are challenges to be overcome, the benefits of collaboration between universities and businesses are clear, and initiatives to facilitate such collaborations are likely to continue to grow in importance in the coming years.

2.2. Integration of Industry and Education

Integration of industry and education is an area of research that focuses on the collaboration between educational institutions and industry to better align education with the needs of the workforce [22]. This approach emphasizes practical training and real-world experiences, which can enhance the employability and job readiness of graduates [23].

Research in this area has identified several key themes that are important for the effective integration of industry and education [24,25]. One key theme is the need for educational institutions to engage with industry partners to develop relevant curricula and training programs. This includes partnerships with businesses, government agencies, and industry associations to ensure that graduates have the necessary skills and knowledge to meet the needs of the job market. Another important theme is the need to provide students with opportunities for experiential learning, such as internships, apprenticeships, and work-integrated learning programs. These opportunities allow students to gain practical experience and develop industry-specific skills, which can improve their employability and make them more attractive to employers [26].

Research has also examined the benefits and challenges of industry and education integration [27,28]. Some of the benefits include improved student employability, better alignment of educational programs with industry needs, and increased innovation and productivity within industry. However, challenges include funding and resource constraints, differences in organizational cultures and priorities, and concerns around academic autonomy and independence [29].

The current research suggests that the integration of industry and education can be an effective way to improve student employability and align education with industry needs [30]. However, effective collaboration between educational institutions and industry partners requires careful planning and coordination, as well as ongoing monitoring and evaluation to ensure that the partnership remains effective and sustainable over time [31].

To further promote the integration of industry and education, many educational institutions have implemented various programs and initiatives aimed at strengthening ties with industry partners [32]. For example, some universities have established industry advisory boards, which provide input into curriculum developments and help to identify opportunities for collaborations with industry. Other institutions have created work-integrated learning programs, which offer students opportunities to gain practical experience through internships, co-op placements, and other forms of experiential learning.

Moreover, some countries have implemented policies and programs aimed at promoting the integration of industry and education at a national level [33]. For instance, Germany has a long-standing tradition of apprenticeships, which allow students to combine vocational training with on-the-job learning, providing a pathway to employment in the manufacturing and engineering sectors. Meanwhile, countries such as Singapore and Australia have established government-funded programs to encourage closer collaboration between educational institutions and industry.

Despite the progress made in this area, challenges still remain. For instance, it can be difficult for educational institutions to maintain effective relationships with industry partners, particularly when there are differences in priorities or goals [34]. There may also be challenges around the sharing of intellectual property and the ownership of research outcomes, which can impact the sustainability of collaborative relationships [35,36].

In conclusion, the integration of industry and education is a rapidly evolving area of research that has the potential to improve student employability and drive economic growth [37,38]. While there are challenges to be overcome, effective collaboration between educational institutions and industry partners can help to ensure that graduates are equipped with the skills and knowledge needed to meet the needs of the workforce [39]. As such, it is likely that initiatives aimed at promoting the integration of industry and education will continue to be an important focus for research and policy in the years to come.

3. Methodology

3.1. Research Design

User satisfaction is a measure that evaluates a customer's experience of the service environment, originating from business management [40,41]. This study uses student satisfaction as an indicator to gauge their actual feelings about the educational services provided by universities and enterprise. University–business cooperation and integration of industry and education are specific manifestations of university–industry collaboration. University–industry collaboration is the central theme that runs through postgraduate education. However, there is a lack of surveys on the satisfaction of professional degree postgraduates.

It is essential to investigate student satisfaction with universities, enterprises, the practice process, and other aspects of engineering practice to gain a better understanding of their perspective as participants in engineering practice activities [42]. Such investigations hold great practical significance for improving the quality of postgraduate education and ensuring that students are satisfied with the services provided.

The survey was conducted using a combination of interviews and online questionnaires. Initially, we visited and conducted research on relevant enterprise practice bases and interviewed teachers and enterprise supervisors. Based on this information, the questionnaire was developed after expert consultation and multiple revisions to ensure that it effectively measured student satisfaction regarding university–industry collaboration.

3.2. Sampling Technique and Measurement Instruments

To ensure the reliability of the research data, we conducted a questionnaire survey in the first half of 2022 using a sample of professional degree graduate students from a university of applied sciences. The survey used stratified random sampling, and 290 valid questionnaires were collected from current graduate students and graduates of different genders and ages. The sample distribution was reasonable, which allowed for a more effective evaluation of the enterprise practice activities and provided the basis for making recommendations.

We developed the "Survey on Professional Degree Graduate Students' Satisfaction with Enterprise Practice" as a measurement tool and revised it after conducting trial research. The questionnaire adopted a self-assessment approach to evaluate the content of enterprise teaching, university teaching, university supervisors, and enterprise supervisors. The questionnaire used a Likert scale for measurement, and after a reliability test, the Cronbach's alpha coefficient of the scale was 0.896, indicating high internal consistency reliability of the scale.

4. Results

As per the research design, the questionnaire data were analyzed using statistical techniques, including analysis of variance and multiple regression analysis.

4.1. Characteristics of the Sample

Table 1 displays initial sample statistics, categorized according to student gender, grade, and age. Of the sample, 153 students were male (52.8%) and 137 were female (47.2%). The distribution of student grades was as follows: 113 1st-year graduate students (39.0%), 77 2nd-year students (26.6%), 50 3rd-year graduate students (17.2%), and 50 graduated postgraduate students (17.2%). As for age distribution, 1 student (0.4%) was under 20 years old, 183 students (63.1%) were between 20 and 25, 94 students (32.4%) were between 26 and 30, and 12 students (4.1%) were over 30.

Table 1. Characteristics of the sample.

Characteristic	Group	*F*	%
Gender	Male	153	52.8
	Female	137	47.2
Grade	1st-year graduate students	113	39
	2nd-year graduate students	77	26.6
	3rd-year graduate students	50	17.2
	Graduated	50	17.2
Age	<20	1	0.4
	20–25	183	63.1
	26–30	94	32.4
	>30	12	4.1

4.2. Descriptive Statistics

The questionnaire comprised 11 questions that evaluated the satisfaction of professional degree students regarding enterprise practice. Results indicated that students were either very satisfied or relatively satisfied with 10 of the aspects (with a mean value ranging from 3.679 to 4.321) (see Table 2). The highest satisfaction level was recorded for the question of "Guidance from university supervisors". However, students were dissatisfied with the relevance of university-based theoretical study and enterprise practice (with a mean value of 2.41). This indicates that while professional degree students generally acknowledge the effectiveness of enterprise practice, the relevance of theoretical learning to enterprise practice requires further enhancement.

University–industry collaboration offers students a realistic corporate environment, complete with engineering projects and implementation sites, wherein they are treated as full-fledged employees of companies and perform practical tasks according to their companies' regulations and management styles [43]. During enterprise practice, students undertake projects as per the directives of their enterprise supervisors. The survey highlights that most students believe enterprise practice aids in enhancing their professional knowledge and practical skills (with a mean value of 4.097).

Table 2. Descriptive statistics.

	Mean	SD	Variance	Kurtosis	Skewness
Enterprise Teaching	3.883	0.915	0.837	0.193	−0.638
University Teaching	4.072	0.77	0.593	0.317	−0.582
Integration of Theory and Practice	4.01	0.83	0.688	0.814	−0.751
Guidance from university supervisors	4.321	0.783	0.613	1.21	−1.067
Guidance from enterprise supervisors	4.079	0.979	0.959	0.834	−1.05
Understanding of the real situation of the future career	4.066	0.771	0.594	−0.085	−0.524
Improvement of professional cognition and practical ability	4.097	0.752	0.565	0.213	−0.603
Contributing to future career planning	4.017	0.742	0.55	−0.014	−0.438
The relevance of major and enterprise practice	3.679	0.879	0.772	−0.422	−0.127
The relevance of university theoretical study and enterprise practice	2.41	0.938	0.879	−0.215	0.364
The relevance of master's thesis and enterprise practice	3.89	0.865	0.749	−0.692	−0.301

University–industry collaboration entails that enterprise practice utilizes a double-supervisor system consisting of on-campus and off-campus guidance [44]. During the enterprise practice period, the off-campus enterprise supervisor assumes the primary supervisory role. They guide the students' practical work in the enterprise, manage the project practice, and provide necessary professional support. Under the double-supervisor system, students are obligated to report their progress in enterprise practice to their on-campus supervisors regularly no less than twice a semester. On-campus supervisors are equipped to provide students with professional theoretical guidance on issues arising during enterprise practice. According to the survey, students expressed greater satisfaction with their university supervisors (mean value 4.321) compared to their enterprise supervisors (mean value 4.079).

As per the university–industry collaboration protocol, students must undertake a practice phase in the enterprise, wherein the enterprise supervisors organize various practical activities such as job skills training, job practice, and involvement in project practice and engineering project management. Unlike theoretical course learning on campus, practical teaching activities occur in the enterprise. The survey reveals that students are relatively dissatisfied with this aspect.

During the enterprise practice stage, students are recommended by their supervisors to join the joint postgraduate training practice base of the university or cooperative enterprises, where their supervisors undertake interdisciplinary research, or the relevant industry management departments. The survey indicates that students have a low recognition of the degree of alignment between the content of enterprise practice and their majors (mean value 3.679). Many students believe that there is still a disconnect between the job arrangement, work content of enterprise practice, and their majors. Thus, relevance and closeness require further improvement.

According to the training program of professional degree postgraduates in universities of applied sciences, on-campus theoretical courses are typically arranged from the second academic year after enrollment after students have completed their engineering practice in enterprises [45]. However, the survey results indicate that more than half of the students feel that there is a disconnect between the theoretical course content taught on campus and the practical experience gained during their enterprise practice (mean value 2.41).

University–industry collaboration requires postgraduates to engage in enterprise practice first, identify problems in practice, search for solutions during theoretical learning, and then test results in enterprise practice [46]. However, according to the survey, students do not consider the relevance of the graduation thesis topic to the content of enterprise practice to be high, with a mean value of 3.89.

4.3. Linear Regression Analysis

The linear regression analysis indicated that postgraduate students evaluated enterprise practices differently depending on their academic year, as shown in Table 3 and Figure 1. Table 3 presents the results of the linear regression analysis, including the standardized coefficients of the model, t-values, VIF values, R^2, adjusted R^2, etc., for the test of the model and the analysis of the formula of the model. The F-test results show a significant difference in the evaluation of enterprise practice among postgraduate students of different academic years (p-value = 0.001), and the original hypothesis that the regression coefficient is 0 is rejected. The VIF was less than 10 for all variables, indicating that the model was well-constructed without multiple cointegration problems. Figure 1 shows the original data plot, the fitted model, and the predicted model values. Overall, the survey statistics achieved the expected objectives and are statistically significant.

Table 3. Linear regression (least squares) analysis of postgraduate students' evaluation of UBC by grade.

	Linear Regression (Least Squares) Analysis $n = 290$								
	Unstandardized Coefficients		Standardized Coefficients	t	p	VIF	R^2	Adjusted R^2	F
	B	Standard Error	Beta						
Constants	2.179	0.492	-	4.434	0.000 ***	-			
Enterprise Teaching	0.218	0.136	0.179	1.598	0.111	3.927			
University Teaching	−0.211	0.118	−0.146	−1.785	0.075 *	2.093			
Integration of Theory and Practice	0.219	0.145	0.163	1.512	0.132	3.634			
Guidance from university supervisors	0.274	0.122	0.193	2.239	0.026 **	2.318			
Guidance from enterprise supervisors	−0.225	0.111	−0.198	−2.031	0.043 **	2.96			
Understanding of the real situation of the future career	−0.005	0.158	−0.003	−0.031	0.975	3.767	0.115	0.077	F = 3 $p = 0.001$ ***
Improvement of professional cognition and practical ability	0.152	0.167	0.102	0.906	0.366	4.001			
Contributing to future career planning	−0.314	0.158	−0.209	−1.989	0.048 **	3.455			
The relevance of major and enterprise practice	−0.322	0.098	−0.254	−3.288	0.001 ***	1.871			
The relevance of university theoretical study and enterprise practice	−0.061	0.07	−0.051	−0.868	0.386	1.101			
The relevance of master's thesis and enterprise practice	0.104	0.087	0.081	1.2	0.231	1.431			
	Dependent variable: Grade								

***, **, * represent 1%, 5%, and 10% level of significance, respectively.

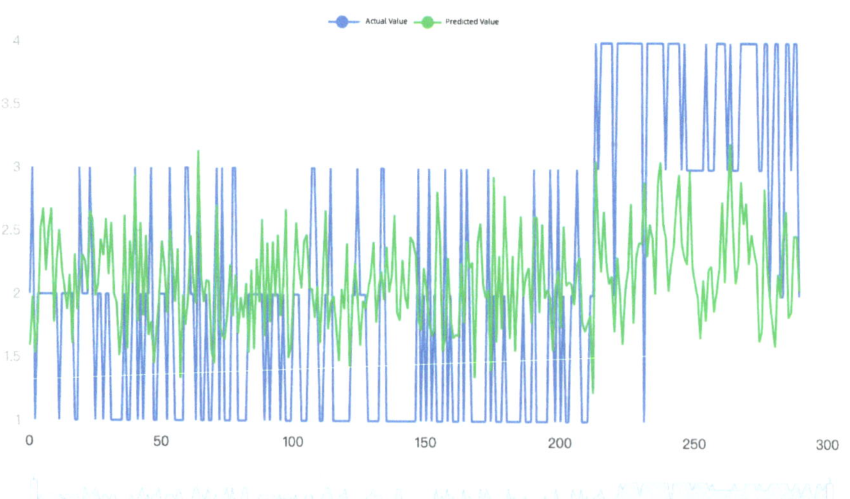

Figure 1. Fitting effects of different grades of postgraduate students on the evaluation of UBC.

5. Discussion

University–industry collaboration offers a realistic work environment, enhancing students' understanding of their future career prospects, improving their professional cognition and practical abilities, and assisting them in career planning [47]. This approach has been widely praised by students, suggesting that the innovative approach to postgraduate education through university–industry collaboration is well-received and supported. In contrast to traditional academic postgraduate education, this mode of education allows students to participate in engineering practice beforehand, fostering problem-solving skills and a teamwork mentality, ultimately aiding students in planning their future careers.

University–industry collaboration offers students an authentic engineering environment wherein they become actively involved and transition from passive to active learners [48]. By identifying practical problems and constructing a system of engineering knowledge with guidance from both enterprise supervisors and university supervisors, students develop the competencies required to become a real-world engineer [49].

The findings of the survey suggest that a certain percentage of students are dissatisfied with the level of integration between theoretical learning and practical content in their enterprise practice. Furthermore, a small proportion of students expressed dissatisfaction with the alignment between their graduation thesis topics and the practical content of their enterprise practice, indicating a potential mismatch between the content of their enterprise practice and the university curriculum.

There are several factors that contribute to the current curriculum's inability to meet the needs of enterprises. Firstly, the curriculum is primarily theoretical and academic, failing to keep pace with timely industrial advancements. Furthermore, the learning methods and practical courses are outdated and disconnected from the professional field, and course content updates are infrequent. Secondly, the practical courses lack systematic design and collaborative development, leading to postgraduates' vocational abilities falling behind industry requirements. University supervisors may not fully appreciate the importance of enterprise practice for postgraduate students, exacerbating the disconnect between theory and practice. Moreover, enterprise supervisors may lack university teaching qualifications, which limits their ability to integrate engineering projects with practical teaching. Postgraduates may face difficulties in management and guidance outside of their busy work, resulting in repetitive work with limited involvement in technical research projects. Consequently, engineering practice may become ineffective and superficial.

6. Conclusions

University–industry collaboration is a novel approach to professional degree postgraduate training for universities of applied sciences. It prioritizes practical training and holds significant exploratory value within the current context of postgraduate education reform in China. To promote the implementation of university–industry collaboration, corresponding measures should be taken.

6.1. Practice-Oriented Professional Training

University–industry collaboration is a practice-oriented educational approach that trains professional degree postgraduate students to meet the unique needs of a country [50]. However, universities of applied sciences that are primarily focused on undergraduate education may have a path dependence on long-standing undergraduate teaching, leading to postgraduate curricula that are merely continuations of undergraduate education [51]. Additionally, supervisors may lack experience in post-graduate education and corporate experience, which makes it challenging to understand the critical attributes of professional degree postgraduate training [52]. To enhance practical and innovative skills, students must stay up to date with the latest advancements in their field and gain practical experience through job attachments, engineering experiences, and research [53]. With the guidance of both enterprise and university supervisors, students can develop into high-level practical professionals capable of identifying problems and devising innovative solutions. The curriculum should incorporate the latest developments in key technologies, industry standards, hot topics, and trends to ensure that students receive a comprehensive education that integrates engineering knowledge and practical ability [54]. This practice-oriented approach fosters a deep understanding of the practical attributes of the profession and effectively applies their skills to real-world problems.

6.2. Engineering Practice throughout the Whole Process of Training

For postgraduate students at universities of applied sciences, developing engineering practice skills is essential [55]. Throughout their education, students can gain industry expertise through various activities. Under the guidance of dual supervisors, postgraduates can select research topics based on industry enterprise needs, solve practical engineering problems, and stay up to date with industry trends and frontier technologies. This can enhance their engineering practice and innovation abilities through their thesis work.

In postgraduate education, university–industry collaboration requires enterprise practice to be closely integrated into the curriculum, and the learning system should be based on engineering cognition and graduation theses to solve practical problems [56]. Students should participate in multi-level engineering practice, combining theory and practice through active inquiry learning, and enhance their engineering practice and innovation abilities. The participation of postgraduates in R&D engineering projects has achieved significant social and economic benefits for enterprises, strongly promoting collaborative innovation between universities and enterprises and reflecting the characteristics of university–industry collaboration in postgraduate education [57].

The specialized courses on campus should focus closely on the entire industrial chain, with an increased proportion of practical teaching [58]. This can be achieved through case-based teaching, engineering problem-oriented teaching, and the use of various teaching methods such as heuristic teaching and on-site teaching. Case teaching can build a bridge between teaching contents and engineering practice, promoting students' understanding and mastery of engineering knowledge. Postgraduates can propose solutions to common and key technical problems in the industry and complete their graduation theses accordingly, creating practical application value with engineering as the background.

6.3. Operation Mechanism of University–Industry Collaboration

To address the issue of the mismatch between the supply of high-level professionals and industrial demands, professional degree graduate students should prioritize university–industry collaboration. This comprehensive approach integrates knowledge transfer, skill development, and value creation, and involves various stakeholders. A robust management system that aligns with educational goals must be established to promote in-depth integration of industry and education. Enterprises are not only places for students to carry out engineering project practice but also becomes important participants in postgraduate education. Universities should make use of the platform of industrial enterprises to transform knowledge achievements and serve the technological innovation needs of industrial enterprises. This integration requires both universities and enterprises to work together in developing a unified body that integrates knowledge transfer, ability cultivation, and value shaping, and to define the relationship of responsibility and rights among universities, enterprises, society, and postgraduates. Strengthening the management and construction of engineering practice bases is crucial for forming a replicable institutional experience. The cooperation with domestic and international enterprises in building postgraduate practice bases should be expanded, and practice bases covering the whole industrial chain of the industry should be established. The relevant management departments of universities should strengthen the selection criteria and assessment indexes of practice bases and establish corresponding assessments, rewards, punishments, and withdrawal mechanisms through standardized management. Policy supports and construction investment should be given to the bases with good cooperation and cultivation results, and enterprises with poor training and low professional fit should be rectified or withdrawn in a timely manner. Third-party social agencies can be introduced to regulate the access, assessment, and withdrawal of enterprises that provide off-campus practice platforms.

6.4. Construction of "Double Supervisor" Faculty

The double-supervisor faculty team consists of university and off-campus enterprise supervisors who jointly guide engineering practice projects, design and implement postgraduate courses, and oversee postgraduate graduation theses. This mechanism for cultivating postgraduates under the double supervisor system [59] can lead to academicization due to the lack of practical experience of university and enterprise supervisors. To address this issue, universities can establish clear requirements and assessment mechanisms [60], such as setting up enterprise project funds and increasing corresponding assessment and title promotion indexes. The selection, guidance, and assessment of enterprise supervisors should also be strengthened, and universities can organize activities such as tutorials for enterprise supervisors to enrich their experience in guiding students [61]. Regular elimination of supervisors is important to ensure their quality [62]. University–industry collaboration in postgraduate education disrupts the traditional theory-practice-theory model and expands the educational significance of postgraduate education [63]. It offers a new model for cultivating professional degree postgraduates and provides valuable insights for other universities of applied sciences.

Author Contributions: Conceptualization and writing original draft, Y.Z.; data curation, Y.Z. and X.C.; review and editing, X.C. All authors have read and agreed to the published version of the manuscript.

Funding: This work was sponsored by the Humanity and Social Science Youth Foundation of the Ministry of Education of China (Grant No. 20YJC880125), the Shanghai Pujiang Program (Grant No. 21PJC063) and Artificial Intelligence Medical Hospital and Locality Cooperation Project in Xuhui District, Shanghai (Grant No. 2021-008).

Institutional Review Board Statement: Not applicable.

Informed Consent Statement: Informed consent was obtained from all subjects involved in the study.

Data Availability Statement: Data can be available from the corresponding author upon request.

Acknowledgments: The authors wish to thank the reviewers for their invaluable comments and suggestions that enhanced the quality of the paper.

Conflicts of Interest: The authors declare no conflict of interest.

References

1. Littenberg-Tobias, J.; Reich, J. Evaluating access, quality, and equity in online learning: A case study of a MOOC-based blended professional degree program. *Internet High. Educ.* **2020**, *47*, 100759. [CrossRef]
2. Barnacle, R.; Dall'Alba, G. Research degrees as professional education? *Stud. High. Educ.* **2011**, *36*, 459–470. [CrossRef]
3. Jones, M. Contemporary trends in professional doctorates. *Stud. High. Educ.* **2018**, *43*, 814–825. [CrossRef]
4. Bedard, K.; Herman, D.A. Who goes to graduate/professional school? The importance of economic fluctuations, undergraduate field, and ability. *Econ. Educ. Rev.* **2008**, *27*, 197–210. [CrossRef]
5. Lewis, E.G.; Cardwell, J.M. The big five personality traits, perfectionism and their association with mental health among UK students on professional degree programmes. *BMC Psychol.* **2020**, *8*, 54. [CrossRef]
6. Hathaway, R.S.; Nagda, B.A.; Gregerman, S.R. The relationship of undergraduate research participation to graduate and professional education pursuit: An empirical study. *J. Coll. Stud. Dev.* **2002**, *43*, 614–631.
7. Anderson, C.; Day, K.; McLaughlin, P. Student perspectives on the dissertation process in a master's degree concerned with professional practice. *Stud. Contin. Educ.* **2008**, *30*, 33–49. [CrossRef]
8. Arranz, N.; Arroyabe, M.F.; Sena, V.; Arranz, C.F.; de Arroyabe, J.C.F. University-enterprise cooperation for the employability of higher education graduates: A social capital approach. *Stud. High. Educ.* **2022**, *47*, 990–999. [CrossRef]
9. Mahfoudh, D.; Boujelbene, Y.; Mathieu, J.P. University-enterprise cooperation: Determinants and impacts. In *Mediterranean Symposium on Enterprise–New Technology Synergy*; Springer: Cham, Switzerland, 2021; pp. 91–121.
10. Orazbayeva, B.; Plewa, C.; Davey, T.; Muros, V.G. The future of University-Business Cooperation: Research and practice priorities. *J. Eng. Technol. Manag.* **2019**, *54*, 67–80. [CrossRef]

11. Alrajhi, A.N.; Aydin, N. Determinants of effective university–business collaboration: Empirical study of Saudi universities. *J. Ind.-Univ. Collab.* **2019**, *1*, 169–180. [CrossRef]
12. Orazbayeva, B.; Davey, T.; Plewa, C.; Galán-Muros, V. Engagement of academics in education-driven university-business cooperation: A motivation-based perspective. *Stud. High. Educ.* **2020**, *45*, 1723–1736. [CrossRef]
13. Bruno, G.; Diglio, A.; Kalinowski, T.B.; Piccolo, C.; Rippa, P. University-Business Cooperation to design a transnational curriculum for energy efficiency operations. *Stud. High. Educ.* **2021**, *46*, 763–781. [CrossRef]
14. Ripoll Feliu, V.; Diaz Rodriguez, A. Knowledge transfer and university-business relations: Current trends in research. *Intang. Cap.* **2017**, *13*, 697–719. [CrossRef]
15. Mascarenhas, C.; Ferreira, J.J.; Marques, C. University–industry cooperation: A systematic literature review and research agenda. *Sci. Public Policy* **2018**, *45*, 708–718. [CrossRef]
16. Clauss, T.; Kesting, T. How businesses should govern knowledge-intensive collaborations with universities: An empirical investigation of university professors. *Ind. Mark. Manag.* **2017**, *62*, 185–198. [CrossRef]
17. Quintana, C.D.D.; Mora, J.G.; Pérez, P.J.; Vila, L.E. Enhancing the development of competencies: The role of UBC. *Eur. J. Educ.* **2016**, *51*, 10–24. [CrossRef]
18. Davey, T.; Plewa, C.; Galán Muros, V. University-business cooperation outcomes and impacts–a European perspective. In *Moderne Konzepte des Organisationalen Marketings*; Springer Gabler: Wiesbaden, Germany, 2014; pp. 161–176.
19. Ranga, M.; Hoareau, C.; Durazzi, N.; Etzkowitz, H.; Marcucci, P.; Usher, A. *Study on University-Business Cooperation in the US*; LSE Enterprise: London, UK, 2013.
20. Gerasimov, B.N.; Vasyaycheva, V.A.; Gerasimov, K.B. Identification of the factors of competitiveness of industrial company based on the module approach. *Entrep. Sustain. Issues* **2018**, *6*, 677. [CrossRef]
21. Rybnicek, R.; Königsgruber, R. What makes industry–university collaboration succeed? A systematic review of the literature. *J. Bus. Econ.* **2019**, *89*, 221–250. [CrossRef]
22. Osadchy, E.A.; Akhmetshin, E.M. Integration of industrial and educational sphere in modernization of economic relations. *J. Appl. Econ. Sci.* **2015**, *10*, 35.
23. Ziatdinov, A.M.; Shaydullina, A.R. Advantages of Information Technologies in Integration of Industry, Science and Education. *Ma a Hao C Aoo oaoo oo a* **2013**, *2*, 98–101.
24. Butt, R.; Siddiqui, H.; Soomro, R.A.; Asad, M.M. Integration of Industrial Revolution 4.0 and IOTs in academia: A state-of-the-art review on the concept of Education 4.0 in Pakistan. *Interact. Technol. Smart Educ.* **2020**, *17*, 337–354. [CrossRef]
25. Moraes, E.B.; Kipper, L.M.; Hackenhaar Kellermann, A.C.; Austria, L.; Leivas, P.; Moraes, J.A.R.; Witczak, M. Integration of Industry 4.0 technologies with Education 4.0: Advantages for improvements in learning. *Interact. Technol. Smart Educ.* **2022**. ahead of print. [CrossRef]
26. Shaydullina, A.R.; Ziatdinov, A.M. Information technologies as the basis for the integration of science, education and industry. *a Aoo oaoo oo a* **2013**, *11*, 172–174.
27. Kartashova, A.; Shirko, T.; Khomenko, I.; Naumova, L. Educational activity of national research universities as a basis for integration of science, education and industry in regional research and educational complexes. *Procedia-Soc. Behav. Sci.* **2015**, *214*, 619–627. [CrossRef]
28. Ren, J.; Wu, Q.; Han, Z.; Wang, D.; Gong, K. Research on the education of industry-education integration for geological majors. *Educ. Sci. Theory Pract.* **2018**, *18*, 1315–1322.
29. Durmuș, A.; Dağli, A. Integration of vocational schools to industry 4.0 by updating curriculum and programs. *Int. J. Multidiscip. Stud. Innov. Technol.* **2017**, *1*, 1–3.
30. Shaidullina, A.R.; Masalimova, A.R.; Vlasova, V.K.; Lisitzina, T.B.; Korzhanova, A.A.; Tzekhanovich, O.M.; Masalimova, A.R. Education, science and manufacture integration models features in continuous professional education system. *Life Sci. J.* **2014**, *11*, 478–485.
31. Bordogna, J.; Fromm, E.; Ernst, E.W. Engineering education: Innovation through integration. *J. Eng. Educ.* **1993**, *82*, 3–8. [CrossRef]
32. Manatos, M.J.; Sarrico, C.S.; Rosa, M.J. The integration of quality management in higher education institutions: A systematic literature review. *Total Qual. Manag. Bus. Excell.* **2017**, *28*, 159–175. [CrossRef]
33. Shaidullina, A.R.; Krylov, D.A.; Sadovaya, V.A.; Yunusova, G.R.; Glebov, S.O.; Masalimova, A.R.; Korshunova, I.V. Model of vocational school, high school and manufacture integration in the regional system of professional education. *Rev. Eur. Stud.* **2015**, *7*, 63. [CrossRef]
34. Ahad, A.; Hasan, S.; Hoque, M.R.; Chowdhury, S.R. Challenges and Impacts of Technology Integration/Up-gradation in the Education Industry: A Case Study. *J. Syst. Integr.* **2018**, *9*, 26–36.
35. Owusu-Manu, D.G.; Edwards, D.J.; Holt, G.D.; Prince, C. Industry and higher education integration: A focus on quantity surveying practice. *Ind. High. Educ.* **2014**, *28*, 27–37. [CrossRef]
36. Shaidullina, A.R.; Sheymardanov, S.F.; Ganieva, Y.N.; Yakovlev, S.A.; Khairullina, E.R.; Biktemirova, M.K.; Kashirina, I.B. The peculiarities of the advanced training of the future specialists for the competitive high-tech industry in the process of integration of education, science and industry. *Mediterr. J. Soc. Sci.* **2015**, *6*, 43. [CrossRef]
37. Watson, P. The role and integration of learning outcomes into the educational process. *Act. Learn. High. Educ.* **2002**, *3*, 205–219. [CrossRef]

38. Zhao, Y.; Rajabov, B.; Yang, Q. Innovation and entrepreneurship education reform of engineering talents in application-oriented universities based on cross-border integration. *Int. J. Front. Eng. Technol.* **2019**, *1*, 56–61.
39. Brekhman, A.I.; Sakhapov, R.L.; Absalyamova, S.G. Innovative model of integration of education and business in the road construction industry. **2014**, *2014*, 41.
40. Gatian, A.W. Is user satisfaction a valid measure of system effectiveness? *Inf. Manag.* **1994**, *26*, 119–131. [CrossRef]
41. Bokhari, R.H. The relationship between system usage and user satisfaction: A meta-analysis. *J. Enterp. Inf. Manag.* **2005**, *18*, 211–234. [CrossRef]
42. Abdurrahaman, D.T.; Owusu, A.; Bakare, A.S. Evaluating factors affecting user satisfaction in university enterprise content management (ECM) systems. *Electron. J. Inf. Syst. Eval.* **2020**, *23*, 1–16. [CrossRef]
43. Orazbayeva, B.; Plewa, C. Academic motivations to engage in university-business cooperation: A fuzzy set analysis. *Stud. High. Educ.* **2022**, *47*, 486–498. [CrossRef]
44. Galan-Muros, V.; Davey, T. The UBC ecosystem: Putting together a comprehensive framework for university-business cooperation. *J. Technol. Transf.* **2019**, *44*, 1311–1346. [CrossRef]
45. Sarpong, D.; AbdRazak, A.; Alexander, E.; Meissner, D. Organizing practices of university, industry and government that facilitate (or impede) the transition to a hybrid triple helix model of innovation. *Technol. Forecast. Soc. Chang.* **2017**, *123*, 142–152. [CrossRef]
46. Shapira, A.; Rosenfeld, Y. Achieving construction innovation through academia-industry cooperation—Keys to success. *J. Prof. Issues Eng. Educ. Pract.* **2011**, *137*, 223–231. [CrossRef]
47. Fadeev, A.S.; Gerdy, V.N.; Baltyan, V.K.; Fedorov, V.G. The Integration of Education, Science, and Industry: The Model of Bauman University. *Vysshee obrazovanie v Rossii = High. Educ. Russ.* **2016**, *4*, 55–63.
48. Abdibekov, U.; Zhakebayev, D.; Karuna, O.; Moisseyeva, Y. Integration of science, education and industry in the republic of Kazakhstan in the context of the development of new educational programs. *Mediterr. J. Soc. Sci.* **2015**, *6*, 364. [CrossRef]
49. Smith, N.M.; Smith, J.M.; Battalora, L.A.; Teschner, B.A. Industry–university partnerships: Engineering education and corporate social responsibility. *J. Prof. Issues Eng. Educ. Pract.* **2018**, *144*, 04018002. [CrossRef]
50. Smirnova, Z.V.; Vaganova, O.I.; Loshkareva, D.A.; Konyaeva, E.A.; Gladkova, M.N. Practice-oriented approach implementation in vocational education. In *IOP Conference Series: Materials Science and Engineering*; IOP Publishing: Bristol, UK, 2019; Volume 483, p. 012003.
51. Rezer, T.; Kuznetsova, E. Practice-oriented training as a mechanism of professional experience development. In *EDULEARN19 Proceedings*; IATED: Arndell Park, Australia, 2019; pp. 2107–2110.
52. Dolgova, V.I.; Belikov, V.A.; Kozhevnikov, M.V. Partnership as a Factor in the Effectiveness of Practice-Oriented Education of Students. *Int. J. Educ. Pract.* **2019**, *7*, 78–87. [CrossRef]
53. Pokholkov, Y.P.; Rozhkova, S.V.; Tolkacheva, K.K. Practice-oriented educational technologies for training engineers. In Proceedings of the IEEE 2013 International Conference on Interactive Collaborative Learning (ICL), Kazan, Russia, 25–27 September 2013; pp. 619–620.
54. Svirin, Y.A.; Titor, S.E.; Petrov, A.A.; Smirnov, E.N.; Morozova, E.A.; Scherbakova, O.Y. Practice-Oriented Model of Professional Education in Russia. *Int. J. Environ. Sci. Educ.* **2016**, *11*, 7368–7380.
55. Sheppard, S.; Colby, A.; Macatangay, K.; Sullivan, W. What is engineering practice? *Int. J. Eng. Educ.* **2007**, *22*, 429.
56. Whitbeck, C. *Ethics in Engineering Practice and Research*; Cambridge University Press: Cambridge, UK, 2011.
57. Cunningham, C.M.; Carlsen, W.S. Teaching engineering practices. *J. Sci. Teach. Educ.* **2014**, *25*, 197–210. [CrossRef]
58. Trevelyan, J. Towards a theoretical framework for engineering practice. *Eng. Pract. A Glob. Context Underst. Tech. Soc.* **2014**, *2014*, 33–60.
59. Slick, S.K. The university supervisor: A disenfranchised outsider. *Teach. Teach. Educ.* **1998**, *14*, 821–834. [CrossRef]
60. Borders, L.D. *The Good Supervisor*; ERIC Clearinghouse: Washington, DC, USA, 1994.
61. Wang, X. The Design and Management of the Supervisor Team for Professional Degree Graduate Students under the School-Enterprise Cooperation Model. In *2015 Conference on Informatization in Education, Management and Business (IEMB-15)*; Atlantis Press: Amsterdam, The Netherlands, 2015; pp. 1033–1037.
62. Tucker, M.K.; Jimmieson, N.L.; Bordia, P. Supervisor support as a double-edged sword: Supervisor emotion management accounts for the buffering and reverse-buffering effects of supervisor support. *Int. J. Stress Manag.* **2018**, *25*, 14. [CrossRef]
63. Jia, G.; Cheng, L.; Wang, S. Application of double supervisor teaching in practical teaching of rehabilitation technology specialty. *Chin. J. Med. Educ. Res.* **2021**, *12*, 668–671.

Article

Engineering Students Education in Sustainability: The Moderating Role of Emotional Intelligence

Teresa Nogueira [1,*] , **Rui Castro** [2] **and José Magano** [3,4]

1   School of Engineering, Polytechnic Institute of Porto (P. Porto), 4249-015 Porto, Portugal
2   INESC-ID/IST, University of Lisbon, 1000-029 Lisboa, Portugal
3   Research Center in Business and Economics (CICEE), Universidade Autónoma de Lisboa, 1150-293 Lisboa, Portugal
4   Higher Institute of Business and Tourism Sciences, Rua de Cedofeita, 285, 4050-180 Porto, Portugal
*   Correspondence: tan@isep.ipp.pt

Abstract: In the context of a lack of quantitative research approaching an engineering education in sustainability, this cross-sectional study aims to investigate whether efforts to promote sustainability education contribute to shaping the beliefs, attitudes, and intentions towards sustainability in a sample of Portuguese engineering schools students; in addition, this study investigates whether emotional intelligence impacts the students' motivation to learn more about sustainability and whether it plays a role in moderating the relationships between those variables. A survey was carried out on a sample of 184 students from two major Portuguese engineering schools. A model was found showing that beliefs, attitudes, and gender are predictors of students' intentions towards sustainability, explaining 62.6% of its variance. Furthermore, the findings reveal that women have stronger beliefs and intentions towards sustainability than men and that students with higher emotional intelligence are more motivated to learn more about sustainability. In addition, emotional intelligence has a negative and significant moderating impact on the relationship between attitudes and students' intentions towards sustainability, being stronger for lower levels of emotional intelligence and having a similar, yet non-significant, effect on the relationship between beliefs and students' intentions towards sustainability. The results suggest that emotional intelligence should be considered a competence and a tool in engineering education in order to enhance students' inclination towards sustainable development.

Keywords: sustainability; emotional intelligence; beliefs; attitude; intention; engineering students; moderation; sustainability education

Citation: Nogueira, T.; Castro, R.; Magano, J. Engineering Students Education in Sustainability: The Moderating Role of Emotional Intelligence. *Sustainability* **2023**, *15*, 5389. https://doi.org/10.3390/su15065389

Academic Editor: Rosabel Roig-Vila

Received: 14 February 2023
Revised: 15 March 2023
Accepted: 16 March 2023
Published: 17 March 2023

1. Introduction

Only in the past few decades have concerns about sustainability and sustainable development become extensively recognized as a societal concern [1]. According to the World Commission on Environment and Development, sustainability is aimed at "promoting harmony among human beings and between humanity and nature" [2]; such "harmony" is translated into the "triple bottom line" concept, which propounds a balance between economic, social, and environmental sustainability [3]. Despite this multidisciplinary perspective of sustainability, there is no consensus about the importance and priorities of each dimension [4]. The sustainable development concept, introduced in the Brundtland Report [2], is referred to as " . . . development that meets the needs of the present without compromising the ability of future generations to meet their own needs"; this has become a reference for scientific research on the environment [5], underlying the progress of our society from a responsible economic perspective, and is in agreement with environmental and natural practices [6]. Sustainability frameworks and approaches for scientific research and environmental management keep evolving [7]. Nevertheless, the terms 'sustainability' and 'sustainable development' are often used as synonyms [8], despite there existing a contradiction in that it is not possible to sustain infinite growth on a limited planet [9]. In this study, we will use the concepts interchangeably, referring to a systems approach that conveys a multidisciplinary sustainability perspective (economic, social, and environmental).

Sustainability has gained significant attention over the years amongst engineers [10], who play a crucial role in designing, building, and maintaining sustainable systems and infrastructure. It is, therefore, of interest to positively influence the attitudes of engineering students towards sustainability; thus, educators and universities should play an important role in fostering their values and beliefs towards sustainability [11]. In fact, education aspires to transform the attitudes and behaviors of forthcoming generations towards sustainability through the combined efforts of educators and educational organizations [11,12]. A person's behavior is determined by how highly a goal is valued and by the degree to which the person expects to succeed [13]. Increasing knowledge could influence one's beliefs, values, and intentions [13,14]. Beliefs convey a person's acceptance that something is true and deal with the establishment of a person's values and convictions, that, in turn, can be transformed by knowledge [15]. Attitudes refer to a person's lasting evaluations, emotional feelings, and inclinations towards some object or idea, and translate a person's beliefs that are then confirmed through actions and thoughts, which influence intentions and thus drive a person's future actions [14]. Ajzen and Fishbein [13] suggest that beliefs drive the creation of attitudes that thus affect the intention of an individual to act.

Educating engineers in sustainability is essential to foster fundamental changes in their conviction and values; given sustainability's transdisciplinary nature [16], such an endeavor could be achieved by inserting educational content into several engineering disciplines, especially those that have substantial roles in achieving sustainability (although the elements necessary to achieve sustainability stem from all aspects of engineering) [17]. Consequently, it is of interest to assess the effectiveness of sustainability educational efforts; this is, indeed, if engineering schools are successful in shaping students' attitudes and beliefs towards sustainability and in developing their sustainability traits. However, there is a lack of sufficient empirical studies that address such a concern [11,18]. Given such a context, we hypothesize the following:

H1. *Beliefs towards sustainability have a significant direct relationship with the intention to act more sustainably.*

H2. *Attitudes towards sustainability have a significant direct relationship with the intention to act more sustainably.*

Another aspect of sustainable development education is that it should consider students' emotional development in order to enhance their skills and academic performance, and lay the foundation for a more collaborative and humane society; thus, building sustainable societies implies developing and managing emotional skills in addition to economic, social, and environmental factors [19]. Emotional intelligence refers to a person's ability to manage their feelings so that those feelings are expressed appropriately and effectively [20], or are referred to as "the individual's ability to use reason to understand and deal with emotions (own and others) and use emotions to understand the context and make more rational decisions" [21]. Previous research has sought to explain how beliefs and attitudes shape favorable behaviors towards the environment [22–25], and has implied that the intelligent usage of emotions is an element of positive environmental behavior [26], namely the ability to assess and regulate one's emotions [26]. On the other hand, emotional intelligence is vital in learning and in individual development [27], contributes to and enhances the cognitive abilities of students [28,29], and is accepted to be essential for the formation of engineers [30]. Emotional intelligence can be seen as a trait, being that the students' emotional profile can vary according to gender and age [31]. In addition, there is evidence of the linkage between sustainable development and emotional intelligence [22,32,33]; this considers the latter to be a dimension of sustainability that plays a principal role in sustainable engineering education. Riemer [34] suggests that emotional intelligence is not only a tool for engineering students while they are learning, but that it also offers career skills for the engineering graduate. Aguilar et al. [26] suggest that considering emotional intelligence as moderating the relationship between beliefs, attitudes, and intentions towards sustainable behavior would contribute to a better understanding and better prediction of sustainable behavior. However, despite the progress that has been made in engineering education towards sustainability, emotional intelligence has not received enough attention and there is a lack of empirical studies that address such a gap [11]. Accordingly, we hypothesize the following:

H3. *Students with higher emotional intelligence will be more motivated to learn more about sustainability.*

H4. *Differences in beliefs, attitudes, and intentions towards sustainability according to gender will be found.*

H5. *Emotional intelligence will play a moderating role in the relationship between attitudes and intentions towards sustainability (H5a), and between beliefs and intentions towards sustainability (H5b).*

The generic research question of this study asks whether sustainability education could generate positive effects in terms of the beliefs, attitudes, and intentions towards sustainability of a sample of engineering students in Portuguese engineering schools; in addition, it asks whether emotional intelligence plays a role in moderating the relationships between these variables and the engineering students' motivation to learn more about sustainability (an exploratory approach that is motivated by the current lack of evidence regarding the role of emotional intelligence in contributing to a sustainable mindset).

As such, this study was carried out on a sample of engineering students in two major Portuguese engineering schools, who undertook courses with basic content on sustainability in the first semester of 2022/23; such content and the effort to deliver it were expected to impact the students' beliefs regarding sustainability, leading to positive changes in their attitudes and intentions, and thus enhancing their sustainability traits. For that purpose, a survey was carried out using measurement instruments drawn from the literature and described in Section 2, namely concerning beliefs, attitudes, intentions towards sustainability and emotional intelligence.

The remainder of the article Is organized as follows: Section 2 describes the procedures, measurement instruments, data analysis method used in the study, and the sample characteristics; Section 3 documents the results, which are discussed in Section 4; and Section 5 presents the conclusions and limitations of the study.

2. Methods

2.1. Procedures

A cross-sectional survey was designed and carried out between 23 November and 19 December 2022 at two Portuguese higher education institutions—the School of Engineering of the Polytechnic Institute of Porto (ISEP-IPP) and the Instituto Superior Técnico of the University of Lisbon (IST-UL)—in order to analyze the impact of sustainability education efforts among engineering students in both schools regarding their beliefs, attitudes, and intentions towards sustainability, and if their emotional intelligence moderated the relationships between those variables. For this purpose, at the beginning of the semester, the authors identified courses for the undergraduate and master's degrees with sustainability content. The identified courses were designed to encourage the teaching of fundamental knowledge on sustainability and sustainable development, enhancing the respective benefits of engineering practice; this includes making decisions concerning materials and processes in project management, stimulating an intrinsic motivation towards sustainability, and, in general, enhancing the students' awareness of sustainability's economic, environmental, and societal dimensions (Elkington's triple bottom line). They were also designed to describe the United Nations Sustainable Development Goals (SDGs), and enhance their significant human-centric attributes [35]. The questionnaire included items adapted from Tang [11] and Rego and Fernandes [36]; authorization from the authors was requested.

The 25 items from Tang [11] were translated from English to Portuguese using the back-translation technique to ensure the quality of the translation. Before starting the fieldwork, a small group of researchers was invited to critique the initial draft of the questionnaire. As a result, the wording and suitability of the form were improved, and the questionnaire was pretested on a group of 31 students. Once making the revisions suggested after this pilot study, the target students were surveyed by administering the questionnaire during classes in the final weeks of the semester. When answering the questionnaire about their beliefs, attitudes, and intentions regarding sustainability, students were asked to compare their position at the time of answering with the one they had at the beginning of the school semester. Any invalid (incomplete or incorrectly completed) responses that were obtained were removed from the analysis. The research protocol included informed consent, which contained the study's objectives and ensured the participants' confidentiality and anonymity.

2.2. Measure Instruments

The questionnaire consisted of three sections. The first section comprised the 23 items of the emotional intelligence scale that was suggested and validated by Rego and Fernandes for the Portuguese population [36]. Prior research applying this instrument reported an internal consistency (Cronbach's alpha) of over 0.7 [37]. Answers were assessed with a 7-point Likert scale (1—'the statement does not apply to me', up to 7—'the statement applies to me completely'). The second section included ordinal-scale items to examine students' beliefs, attitudes, and intentions towards sustainability. These unidimensional variables were measured with five-point Likert scales (1—'strongly disagree' to 7—'strongly agree'), comprising 25 items altogether (Beliefs—6 items; Attitudes—13 items; Intentions—6 items), suggested by Tang [11] in order to gauge the three domains in the context of sustainability education. The third section comprised a sociodemographic questionnaire with questions related to gender (masculine—1; feminine—0;), age, and nationality (Portuguese—1; Other—0), a question to describe the course the respondents were enrolled in ('What course are you currently taking?') and its degree ('Degree of the course you are currently attending?', being 1—Bachelor and 0—Master's degree), a question about the respondents' prior training in sustainability ('Have you had training on sustainability in any course that you attended before the current semester?', being 1—'Yes' and 0—'No'), and, finally, a question on the motivation to acquire more training on the subject ('Do you feel motivated to learn more about sustainability?', being 1—'Yes' and 0—'No').

2.3. Data Analysis

Statistical analyses were carried out using SPSS (version 28.0). The sample was characterized by descriptive analysis (means, standard deviations, percentages, cumulative percentages). The normality of the items' distribution was evaluated by determining skewness and kurtosis indicators: skewness values under 3 and kurtosis values under 10 suggest normality [38]. In order to establish whether items fit the study's data, correlations between items and a principal component analysis (with varimax orthogonal rotation) were carried out for beliefs, attitude, and intentions towards sustainability instruments, which were not validated for the Portuguese population [39]. As such, items with no Pearson's r above 0.3 were removed, as well as items with factor loadings below 0.5 [40]. To assess the construct's reliability, Cronbach's alpha coefficients were determined ($\alpha > 0.7$) [41]. Convergent validity was evaluated by determining each construct's composite reliability ($CR > 0.7$) and the average variance extracted ($AVE > 0.5$), applying Fornell and Larcker's cut-off values [42]. Discriminant validity was assessed by ensuring that shared variance among the variables did not exceed the square root of the AVE. A hierarchical multiple regression was performed to estimate the effects of the control variables and attitudes and beliefs towards sustainability on students' intentions towards sustainability. Furthermore, differences were investigated using the independent t-test and the Mann–Whitney test, whose interpretation followed Cohen's [43] guidelines. The statistical significance threshold was set at 0.05. Finally, the PROCESS macro for SPSS [44] was applied to examine whether emotional intelligence would moderate the relationship between the beliefs, attitudes, and intentions to act towards sustainability.

2.4. Sample

The fieldwork yielded a sample of 184 valid responses. The general characteristics of this sample are reported in Table 1. As can be seen, most students were males (72.8%), whereas the number of students from both institutions was balanced (ISEP-IPP—48.4 %; IST-UL—51.6%). Over two thirds were master's students (first year), whereas 31.0% were undergraduate students (last year).

Nearly 75% of the sample was made of Portuguese students, whereas 26.1% had other nationalities (especially associated with master's degrees of IST-UL); the most represented nationalities were Indian (10 students), German (6), French (5), Spanish (4), and Italian (4) (Figure 1).

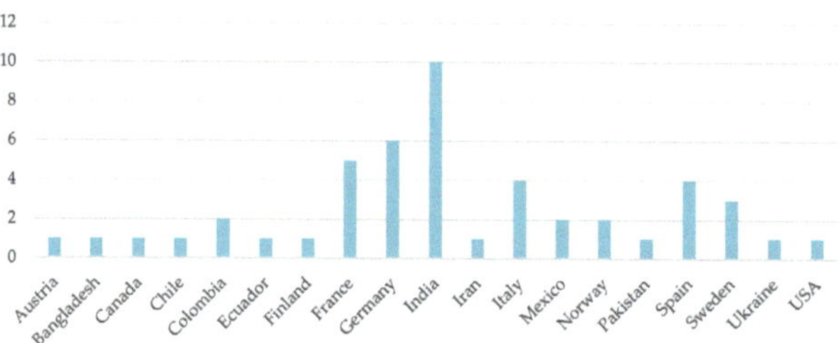

Figure 1. Non-Portuguese students—country of origin.

Most students were enrolled in Electrical and Computer Engineering, Energy Engineering and Management, and Power Systems Engineering (Figure 2); other courses (8%) included Engineering and Industrial Management, Biomedical Engineering, Energy Technologies, and Mechanical Engineering.

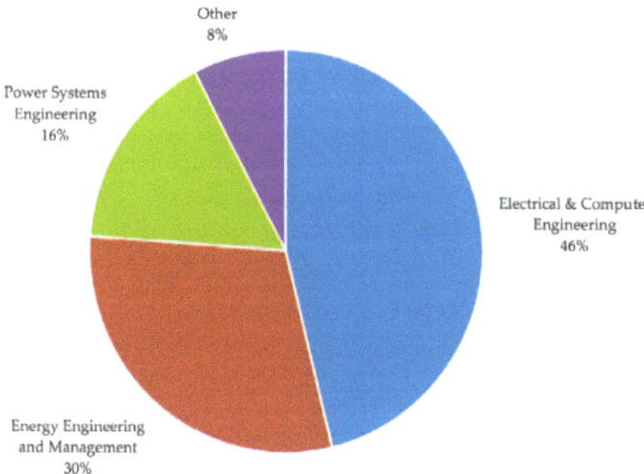

Figure 2. Sample students' engineering courses.

Table 1. General characteristics of the sample.

Variables	Total			ISEP-IPP			IST-UL		
	N	Percent	Cumulative Percent	N	Percent	Cumulative Percent	N	Percent	Cumulative Percent
Sample	184	100	100	89	48.4		95	51.6	
Gender									
Male	134	72.8	72.8	71	38.6	38.6	63	34.2	34.2
Female	50	27.2	100.0	18	9.8	48.4	32	17.4	51.6
Degree									
Undergraduates	57	31.0	31.0	55	29.9	29.9	2	1.1	1.1
Master's	127	69.0	100.0	34	18.5	48.4	93	50.5	51.6
Nationality									
Portuguese	136	73.9	73.9	85	46.2	46.2	51	27.7	27.7
Other	48	26.1	100.0	4	2.2	48.4	44	23.9	51.6
	M	SD	Min–Max	M	SD	Min–Max	M	SD	Min–Max
Age	24.0	4.882	19–55	24.6	6.410	19–55	23.5	2.720	20–37

Note: N = frequencies; % = percentage; M = mean; SD = standard deviation; Min = minimum; Max = maximum.

3. Results

Table 2 presents the descriptive statistics related to the items of the instruments used in this study: Beliefs, Attitudes, Intentions, and Emotional intelligence. All skewness and kurtosis values are within the normative values, ensuring the normality of the distribution.

Table 2. Item frequencies.

	Variables/Items	Scale	*M*	*SD*	Skewness	Kurtosis
	Beliefs	1–5	3.94	0.744	−0.516	0.633
BF01	I feel more morally obliged to do something about environmental problems.		3.99	0.890	−0.647	0.181
BF02	I feel more morally obliged to do something about social problems.		3.84	0.903	−0.528	0.014
BF03	I think I should take more responsibility for sustainable development.		3.98	0.890	−0.720	0.359
BF04	I believe that humans have the right to subdue and control nature.		3.46	1.270	−0.408	−0.885
BF05	I believe that humans should adapt to nature rather than modify it to suit them.		3.93	1.051	−0.750	−0.160
BL06	I think it is important to control human population to ensure social sustainability.		3.46	1.157	−0.412	−0.580
	Attitudes	1–5	3.82	0.565	−0.556	0.674
AT01	I am more aware of current environmental, social, economic, and cultural issues.		4.14	0.738	−1.294	3.565
AT02	I can analyze issues related to sustainable development more holistically.		3.92	0.731	−0.636	1.152
AT03	I am more concerned about environmental pollutions.		4.09	0.854	−0.924	1.014
AT04	I am more willing to safeguard sustainable development.		4.10	0.721	−0.689	1.266
AT05	I make an effort to use green products and services whenever possible.		3.76	0.899	−0.682	0.416
AT06	I refuse the use of packaging.		2.96	1.073	−0.085	−0.638
AT07	I set aside garbage for reuse, recycling, or safe disposal.		4.15	0.969	−1.185	0.994
AT08	I reduce the use of air-conditioning, lighting, and domestic electrical appliances.		3.79	1.077	−0.695	−0.076
AT09	I consciously make a change in my lifestyle to reduce my carbon footprint.		3.48	1.040	−0.486	−0.035
AT10	I consciously reduce the amount of waste generated from my daily activities.		3.67	0.965	−0.522	−0.052
AT11	I make an effort to use energy and resources more efficiently.		4.09	0.847	−1.049	1.427
AT12	I reduce water consumption.		3.91	0.916	−0.636	0.087
AT13	I am willing to pay more for energy-efficient products.		3.60	1.087	−0.641	−0.108
	Intentions	1–5	4.02	0.680	−0.794	1.716
IT01	I prefer to work for an environmentally responsible employer in the future.		4.17	0.836	−1.065	1.755
IT02	I prefer to work for a socially responsible employer in the future.		4.22	0.789	−1.229	2.812
IT03	I intend to change/continue to change my lifestyle for better sustainability.		4.22	0.746	−0.946	1.535
IT04	I will promote the concept of sustainable development to my family and friends.		4.13	0.792	−0.637	0.277
IT05	I will participate in campaigns/causes that promote sustainable development.		3.57	1.006	−0.359	−0.187
IT06	I will apply the concept of triple bottom line more in making decisions.		3.80	0.890	−0.498	0.268
	Emotional intelligence	1–7	5.39	0.629	−0.094	−0.165
EI01	When a friend of mine wins an award, I feel happy for him.		6.35	0.875	−1.350	1.331
EI02	I am indifferent to the happiness of others. (i)		5.58	1.719	−1.203	0.384
EI03	I feel good when a friend of mine gets a compliment.		5.97	1.123	−1.443	3.328
EI04	I live my friends' problems as if they were my problems.		4.17	1.467	−0.054	−0.439
EI05	I get irritated when people criticize me—even though I know other people are right. (i)		4.68	1.514	−0.330	−0.529
EI06	It is hard for me to accept a criticism. (i)		5.29	1.267	−0.590	−0.148
EI07	I don't deal well with the criticisms they make of me. (i)		5.36	1.255	−0.740	0.304
EI08	I have difficulty talking to people who do not share the same views as mine. (i)		5.52	1.297	−0.909	0.383
EI09	When I'm defeated in a game, I lose control. (i)		6.06	1.335	−1.892	3.660
EI10	I can remain calm even when others are angry.		5.11	1.562	−0.791	0.002
EI11	I react calmly when I am under stress.		4.48	1.578	−0.166	−0.654
EI12	Am I really able to control my own emotions?		4.81	1.426	−0.368	−0.389
EI13	I do my best to achieve the goals I set for myself.		5.73	1.310	−0.970	0.335
EI14	I usually encourage myself to do my best		5.75	1.303	−1.055	0.525
EI15	In general, I usually set goals for myself.		5.63	1.454	−1.049	0.327
EI16	I know well what I feel.		5.34	1.305	−0.583	−0.078
EI17	In general, I am aware of my feelings.		5.58	1.286	−1.109	1.735
EI18	I understand the causes of my emotions.		5.27	1.273	−0.540	0.051
EI19	I understand my feelings and emotions.		5.29	1.271	−0.524	−0.241
EI20	When I'm sad, I know what the reasons are.		5.15	1.410	−0.593	−0.397
EI21	I try to understand the feelings of the person I am listening to.		5.78	1.124	−0.915	0.744
EI22	I can understand my friends' emotions and feelings by seeing their behavior.		5.44	1.134	−0.544	0.034
EI23	I strive to understand other people's points of view.		5.58	1.269	−1.133	1.637

Note: *M* = mean; *SD* = standard deviation; (i) reverse item.

Several items were removed from subsequent analysis: items BF4, BF5, and BF6 displayed weak correlations (r < 0.3); items AT5, AT6, AT7, and AT13 had low factor loadings (<0.5). The instruments' internal consistency was assessed using Cronbach's alpha (Table 3). The results found that Cronbach's α of each construct was greater than

0.7 (0.78–0.89), showing high reliability for our survey instrument [45]. Furthermore, as shown in Table 3, the factor loadings of all the constructs exceeded 0.5 and thus conformed to the test of item reliability [40]. In addition, the composite reliabilities of all constructs exceeded the 0.7 (0.76–0.94) cut-off value, as recommended by Fornell and Larcker [42]. In addition, the average variance extracted from each construct exceeded 0.5, indicating convergent validity [42]. In short, the convergent validity test demonstrated that the proposed constructs were adequate.

Table 3. Variables descriptive statistics and reliability.

Construct	Factor Loadings	*CR*	*AVE*	Cronbach's α
Beliefs		0.756	0.509	0.890
BF01	0.746			
BF02	0.714			
BF03	0.678			
Attitudes		0.913	0.539	0.778
AT01	0.843			
AT02	0.821			
AT03	0.690			
AT04	0.706			
AT08	0.694			
AT09	0.625			
AT10	0.777			
AT11	0.674			
AT12	0.751			
Intentions		0.864	0.516	0.837
IT01	0.793			
IT02	0.697			
IT03	0.710			
IT04	0.730			
IT05	0.687			
IT06	0.687			

Note: *CR* = composite reliability; *AVE* = average variance extracted.

In addition, discriminant validity was employed to measure how much the constructs differed; if the items in a construct were more strongly associated with each other than with the items measuring other constructs, the measure was seen as having discriminant validity. As portrayed in Table 4, the shared variance among the variables did not exceed the square root of the AVE. Therefore, discriminant validity is confirmed.

Table 4. Construct means, standard deviations, correlations, and AVE.

Variable	M	SD	BLF	ATT	INT	AVE
Belief	3.94	0.74	**0.71**			0.454
Attitude	3.88	0.58	0.65 **	**0.73**		0.488
Intention	4.02	0.68	0.66 **	0.69**	**0.72**	0.563

Note: ** $p < 0.01$. BLF—Beliefs; ATT—Attitudes; INT—Intentions; AVE = average variance extracted; bold (diagonal) = AVE square roots.

A hierarchical multiple regression was performed to estimate the effects of the control variables and attitudes and beliefs towards sustainability on students' intentions towards sustainability. An independence of residuals was found, as assessed by a Durbin–Watson statistic of 1.946. The independent variables were entered in three blocks; consequently, two nested models were generated. Model 1 estimated the effect of the control variables (gender, age and country of origin were not statistically significant); Model 2 added beliefs, and Model 3 added attitudes (Table 5). The results revealed that attitudes and beliefs towards sustainability have a statistically significant positive direct effect on intentions towards sustainability. The full model of gender, beliefs and attitudes that was used to predict intentions towards sustainability (Model 3) was statistically significant, $R^2 = 0.626$, $F(3,180) = 100.58$, $p < 0.001$; adjusted $R^2 = 0.620$. The addition of beliefs to the prediction of intentions towards sustainability (Model 2) led to a statistically significant increase in R^2 of 0.394, $F(1, 181) = 125.741$, $p < 0.001$. The addition of attitudes to the prediction of intentions towards sustainability (Model 3) also led to a statistically significant increase in R^2 of 0.193, $F(1, 180) = 93.091$, $p < 0.05$. Therefore, H1 and H2 are accepted.

Table 5. Hierarchical multiple regression for intentions towards sustainability.

Variables	Model 1			Model 2			Model 3		
	B	Std β	t-Value	B	Std β	t-Value	B	Std β	t-Value
(Constant)	4.24 **		44.85	1.78 **		7.70	0.58 *		2.57
Gender	−0.30 **	−0.20	−2.73	−0.10	−2.06	−1.11	−1.60 *	−0.11	−2.24
Beliefs				0.59 **	0.64	11.21	0.24 **	0.26	4.24
Attitudes							0.68 **	0.58	9.65
R^2	0.039			0.433			0.626		
F	7.44 *			69.14 **			100.58 **		
ΔR^2	0.039			0.394			0.193		
ΔF	7.44 *			7.44 **			93.09 **		

Note: R^2 = explained variance; B = shared variance between variables; Std β = standardized regression coefficient; t = Student's t-test. * $p < 0.05$; ** $p < 0.001$.

A Mann–Whitney test was conducted to compare beliefs, attitudes, and intentions towards sustainability for gender, for the groups who have/have not had prior training in sustainability, school of origin, and nationality; differences were found only for gender (Table 6) concerning beliefs and intentions towards sustainability, being that, in both cases, females scored significantly higher. Consequently, H4 is accepted. An independent-samples t-test was run to determine whether there were differences in emotional intelligence between men and women; women scored higher in EI ($M = 5.43$, $SD = 0.49$) than men ($M = 5.37$, $SD = 0.67$), but this difference was not statistically significant, $M = −0.23$, 95% CI [−0.50, 0.04], $t(182) = −1.688$, $p = 0.093$. Another independent-samples t-test was run to determine whether there were differences between the emotional intelligence levels of the respondents who felt more motivated to learn about sustainability and those who did not; students that felt more motivated scored higher in EI ($M = 5.30$, $SD = 0.89$) than men ($M = 5.12$, $SD = 0.78$), making this difference statistically significant, $M = −0.18$, 95% CI [−0.42, 0.06], $t(182) = −1.456$, $p = 0.147$, and thus supporting H3.

Table 6. Differences by gender and motivation to learn more about sustainability.

Variable	Gender	N	MR	Md	U	Z	p	r
Beliefs	Male	134	85.18	4.00	2369.5	−3.09	**0.002**	−0.23
	Female	50	112.11	4.33				
Attitudes	Male	134	90.47	3.90	3078.0	−0.848	0.396	−0.06
	Female	50	97.94	3.90				
Intentions	Male	134	86.37	4.00	2528.5	−2.568	**0.010**	−0.19
	Female	50	108.93	4.17				

Table 6. *Cont.*

Variable								
	Gender	*N*	*M*	*SD*	*t*	*df*	*p*	*d*
Emotional intelligence	Male	134	5.37	0.67	−0.670	119	0.504	0.63
	Female	50	5.43	0.49				
	Motivation to Learn More about Sustainability	*N*	*M*	*SD*	*t*	*df*	*p*	*d*
	No	102	5.26	0.62	−3.051	182	**0.003**	−0.61
	Yes	82	5.54	0.61				

Note: N = frequencies; MR = mean rank; Md = median; U = Mann–Whitney test; r = effect size; M = mean; SD = standard deviation; t = t-test; df = degrees of freedom; p = p-value; d = Cohen's d. In bold: statistically significant values.

A moderation analysis was conducted to test the hypothesis that emotional intelligence plays a moderating role in the relationship between attitudes (H5a) and beliefs (H5b), and intentions towards sustainability. Firstly, intentions towards sustainability was set as the dependent variable, attitudes was set as the independent variable, and emotional intelligence was set as the moderating variable. The results indicate the negative and significant moderating effect of EI on the relationship between ATT and INT ($B = -0.224$, 95% CI (-0.39; -0.06), $t = -2.717$, $p = 0.007$), being that the interaction term increases the explained variance of INT ($\Delta R^2 = 1.7\%$, ($F(1,180) = 7.38$, $p = 0.007$), supporting H5a. A simple slope analysis carried out to better understand the nature of this moderating effect, at the mean of EI and ± 1 standard deviation from the mean, reveals that the impact of ATT on INT is stronger for lower levels of EI ($B = -0.926$, 95% CI (0.77; 1.00), $t = 11.859$, $p < 0.001$) than for higher EI ($B = -0.644$, 95% CI (0.47; 0.82), $t = 7.422$, $p < 0.001$); in other words, as the level of EI increases, the strength of the relationship between ATT and INT decreases. Similarly, students' intentions towards sustainability was again set as the dependent variable, beliefs was set as the independent variable, and emotional intelligence was set as the moderating variable. The results indicate the negative yet non-significant moderating impact of EI on the relationship between BLF and INT ($B = -0.096$, 95% CI (-0.23; -0.04), $t = -1.405$, $p = 0.162$), increasing the explained variance of INT ($\Delta R^2 = 0.6\%$ ($F(1,180) = 1.97$, $p = 0.162$); thus, H5b is rejected. In sum, all hypotheses were accepted except H5b.

4. Discussion

The results document that, after being delivered, the educational sustainability content in the courses undertaken by respondents revealed a widespread moderate to strong agreement between beliefs, attitudes, and intentions towards sustainability, which is in line with [11,12,46]. Among the items that scored the highest within beliefs, one can highlight the conviction that students should act with regard to environmental issues and should take responsibility for sustainable development. Regarding attitudes towards sustainability, they indicate a willingness to set aside garbage for reuse, recycling, or safe disposal, and show a stronger awareness of current environmental, social, economic, and cultural issues. These results align with the literature, which confirms that sustainability education can positively affect students' ecological footprint [18]. The item that scored the lowest was refusing the use of packaging ($M = 2.96$, $SD = 1.073$), possibly reflecting an impediment to the acceptance of a more sustainable lifestyle. In terms of the intention to act towards sustainability, the results enhance the students' willingness to work with environmentally and socially responsible employers and their inclination to adapt their lifestyle for higher levels of sustainability. In general, the courses have created motivation among the students to follow a sustainable lifestyle and practices, as can be inferred by the mean ratings obtained in most items (all above 3.0, except for item AT6).

A model was found in which attitudes, beliefs, and gender significantly correlate with students' intentions towards sustainability; as predictors of such an intention, they explain, as a whole, 62.6% of the variance in these dependent variables. This is in line with previous research [11,18] that has suggested that sustainability education could change beliefs and attitudes, and, in turn, impact students' intentions to act towards sustainability; this is also in line with the suggestion that there may be differences concerning gender [31]. Furthermore, women scored significantly higher in attitudes and intentions towards sustainability, which is possibly linked to the fact that they also revealed higher levels of global emotional intelligence; this is an outcome that aligned with Zhoc et al. [29], who state that EI contributes to key learning outcomes in higher education (including social, cognitive and self-growth outcomes) and that women develop higher levels of emotional intelligence [30]. However, this is in contrast with the results obtained by Ryu et al. [18], who found that gender had no statistical significance in post-sustainability training regarding attitudes and intentions towards sustainability.

Several studies report that emotional intelligence has positive effects on engineering learning [32,37] and that it enhances behaviors towards sustainability [22,26]; this study's results are similar, finding that students with higher emotional intelligence are more motivated to learn about sustainable development and how they can help build a more sustainable future for all. On the other hand, EI was found to negatively and significantly moderate the relationship between attitudes and students' intentions towards sustainability, being that the lower the emotional intelligence, the stronger the relationship. A possible explanation for such an outcome is that the higher the emotional intelligence, the less conditioned students are by beliefs or attitudes, likely because they are more complex and may reflect more on sustainability issues. However, the moderating effect of EI on the relationship between beliefs and intentions towards sustainability in our sample was not statistically significant, though it was not distant from the effect on the relationship between attitudes and intentions towards sustainability.

The findings of this study underline the impact of inserting educational content on sustainability into engineering courses, which may, among engineering students, generate stronger beliefs, attitudes, and intentions to act more sustainably and also enhance the effect of emotional intelligence in the learning process. Among the reasons to include emotional intelligence in an education on sustainability in engineering, one can mention that it could improve the ability of engineers to understand the emotional context of sustainability issues and to communicate effectively with stakeholders (for example, engineers who are trained in emotional intelligence are better equipped to understand the emotional barriers that may prevent stakeholders from engaging in sustainability initiatives). In addition, emotional intelligence helps engineers recognize the impact of their emotions and biases on the sustainability decision-making process and take steps to mitigate these effects. In addition, emotional intelligence enables engineers to collaborate effectively with others in promoting sustainability by fostering trust, empathy, and mutual respect. However, most of all, emotional intelligence impacts the learning process itself [20,34,47], suggesting that it should be adequately used to ensure that education on sustainability is more effective among engineering students; as suggested by some authors, EI could even be seen as a competence for engineering education [30,48]. In that regard, a recommendation that may contribute to the education of engineers towards sustainability is integrating EI-related skills into engineering curricula to foster its relevance in education, across disciplines, and in society. Furthermore, improvements in EI may support students to build up knowledge in their discipline more thoroughly, alongside other core skills needed for becoming an engineer [30]. That could be achieved, for instance, by using EI-oriented contents, context-specific role-plays, PBL, or exercises that enhance context and self-awareness, communication skills, team working, the conveyance of ideas, the acceptance of criticism, learning to adapt, leadership [30,34,37,49], and reflection skills and abilities [34]; this is compared to embodying sustainability topics (e.g., sustainable engineering, sustainable technologies and processes, risk and sustainable analysis, sustainable engineering design, and leadership, as recommended by Boyle [17]).

Our results and the implications described above should take into account that this study has some limitations. Firstly, the sample could benefit from being larger and including a more diverse range of students, namely, to allow for investigating significant differences in engineering disciplines and courses. Another limitation is that we relied on the assumption that all the students provided honest answers to the questionnaire; because the respondents were asked to answer it at the end of a lesson, we should expect that some may not have taken it seriously. Furthermore, being a cross-sectional study, we tried to capture whether the educational content on sustainability that was included in the semester's courses indeed impacted the students; nevertheless, we had to rely on the respondents' judgment regarding whether that content had changed their beliefs, attitudes, and intentions to act more sustainably from the beginning to the end of the semester. Another shortfall is that we only approached engineering students on their way to completing undergraduate courses or attending master's degree lectures. The reason for this decision was the consideration that sustainable engineering demands the capability to realize the multifaceted systems that exist within the environment and in society, as well as the restraints on those systems, and thus a greater maturity than that of most traditional engineering disciplines [17]. However, EI competencies also vary with age, suggesting that their use as an effective sustainability learning tool should also involve students at earlier stages of their undergraduate education, as there is also evidence that we can influence students' sustainable behavior then as well [50]. A longitudinal study may be of interest to ensure that the same cohort of students is followed over time and that more reliable conclusions are drawn concerning the impact of efforts to provide education on sustainability and the use of EI within such a context. Finally, our study has addressed emotional intelligence as a whole; that is, we have not decomposed EI into a series of sub-dimensions that deserve analysis on their own (namely, attention to one's emotions, sensitivity to others' emotions, emotional maturity, empathy and emotional contagion, understanding of the causes of one's emotions, self-encouragement, understanding of one's emotions, and emotional self-control). Students could score differently in such sub-dimensions, possibly affecting the effectiveness of the educational approach in student groups, namely concerning gender and age [31].

5. Conclusions

This study was conducted to examine the relationship between engineering students' beliefs and attitudes and also between attitudes and intentions; furthermore, this was to investigate whether emotional intelligence acts as a moderating variable in those relationships, thus contributing to the literature, which lacks quantitative studies that address sustainability in engineering education and the consideration of emotional intelligence in the learning process. The results indicated that the engineering students became more aware of sustainability and strengthened their beliefs, attitudes, and intentions to act towards it in the future after undertaking courses with educational content on sustainability; however, in students with higher EI, the relationships between those variables weaken. In addition, students with higher global emotional intelligence were more willing to learn about sustainability. Such findings support the inclusion of emotional intelligence as a moderator variable in the relationship between engineering students' beliefs, attitudes, and intentions towards; as such, sustainability could help one understand and improve the models used for predicting sustainable behavior, as well as engineering curricula and activities used to enhance effective sustainability learning and graduates' skills and abilities. Emotional intelligence could be an important factor in learning about sustainability, as it helps individuals develop a greater appreciation for the interconnectedness of people, nature, and the environment; EI is known to allow individuals to build strong relationships with others, to be more resilient, flexible, and open-minded in their approach to sustainability, and to be more proactive in their efforts to address sustainability challenges. The ultimate goal would be to form engineers that are capable of addressing those challenges innovatively and holistically; by fostering emotional intelligence in individuals, we can help build a more sustainable future for all.

Author Contributions: Conceptualization, T.N. and J.M.; methodology, T.N. and J.M.; formal analysis, T.N. and J.M.; investigation, T.N., R.C. and J.M.; writing—original draft preparation, T.N. and J.M.; writing—review and editing, T.N., R.C. and J.M. All authors have read and agreed to the published version of the manuscript.

Funding: This research was funded by national funds through FCT—Fundação para a Ciência e a Tecnologia, under the projects UIDB/50021/2020.

Institutional Review Board Statement: Not applicable.

Informed Consent Statement: Informed consent was obtained from all subjects involved in the study.

Data Availability Statement: Datasets are available upon request to the authors.

Conflicts of Interest: The authors declare no conflict of interest.

References

1. Dyllick, T.; Hockerts, K. Beyond the business case for corporate sustainability. *Bus. Strategy Environ.* **2002**, *11*, 130–141. [CrossRef]
2. WCED. *Our Common Future: World Commission on Environment and Development*; Oxford University Press: Oxford, UK, 1987; pp. 1–91.
3. Elkington, J. Towards the sustainable corporation: Win-win-win business strategies for sustainable development. *Calif. Manag. Rev.* **1994**, *36*, 90–100. [CrossRef]
4. Whyte, P.; Lamberton, G. Conceptualising sustainability using a cognitive mapping method. *Sustainability* **2020**, *12*, 1977. [CrossRef]
5. Alvarado-Herrera, A.; Bigne, E.; Aldas-Manzano, J.; Curras-Perez, R. A scale for measuring consumer perceptions of corporate social responsibility following the sustainable development paradigm. *J. Bus. Ethics* **2017**, *140*, 243–262. [CrossRef]
6. Glavič, P.; Lukman, R. Review of sustainability terms and their definitions. *J. Clean. Prod.* **2007**, *15*, 1875–1885. [CrossRef]
7. Ruggerio, C.A. Sustainability and sustainable development: A review of principles and definitions. *Sci. Total Environ.* **2021**, *786*, 147481. [CrossRef]
8. Olawumi, T.O.; Chan, D.W. A scientometric review of global research on sustainability and sustainable development. *J. Clean. Prod.* **2018**, *183*, 231–250. [CrossRef]
9. Spaiser, V.; Ranganathan, S.; Swain, R.B.; Sumpter, D.J. The sustainable development oxymoron: Quantifying and modelling the incompatibility of sustainable development goals. *Int. J. Sustain. Dev. World Ecol.* **2017**, *24*, 457–470. [CrossRef]
10. Rosen, M.A. Engineering and sustainability: Attitudes and actions. *Sustainability* **2013**, *5*, 372–386. [CrossRef]
11. Tang, K.H.D. Correlation between sustainability education and engineering students' attitudes towards sustainability. *Int. J. Sustain. High. Educ.* **2018**, *19*, 459–472. [CrossRef]
12. Andersson, K.; Jagers, S.C.; Lindskog, A.; Martinsson, J. Learning for the future? Effects of education for sustainable development (ESD) on teacher education students. *Sustainability* **2013**, *5*, 5135–5152. [CrossRef]
13. Ajzen, I.; Fishbein, M. *Understanding Attitudes and Predicting Social Behavior*; Prentice-Hall: Hoboken, New Jersey, USA, 1980.
14. Perloff, R.M. *The Dynamics of Persuasion: Communication and Attitudes in the 21st Century*, 2nd ed.; Routledge: New York, NY, USA, 2007. [CrossRef]
15. Wyer, R.S., Jr.; Albarracin, D. Belief formation, organization, and change: Cognitive and motivational influences. In *The handbook of Attitudes*; Albarracin, D., Johnson, B.T., Zanna, M.P., Eds.; Lawrence Erlbaum Associates Publishers: Mahwah, NJ, USA, 2005; pp. 273–322.
16. Tejedor, G.; Segalàs, J.; Rosas-Casals, M. Transdisciplinarity in higher education for sustainability: How discourses are approached in engineering education. *J. Clean. Prod.* **2018**, *175*, 29–37. [CrossRef]
17. Boyle, C. Considerations on educating engineers in sustainability. *Int. J. Sustain. High. Educ.* **2004**, *5*, 147–155. [CrossRef]
18. Ryu, H.C.; Brody, S.D. Examining the impacts of a graduate course on sustainable development using ecological footprint analysis. *Int. J. Sustain. High. Educ.* **2006**, *7*, 158–175. [CrossRef]
19. Estrada, M.; Monferrer, D.; Rodríguez, A.; Moliner, M.Á. Does emotional intelligence influence academic performance? The role of compassion and engagement in education for sustainable development. *Sustainability* **2021**, *13*, 1721. [CrossRef]
20. Goleman, D. *Emotional Development and Emotional Intelligence: Educational Implications*; Basic Books: New York, NY, USA, 1997.
21. Salovey, P.; Mayer, J.D. *Emotional Intelligence*; Dude Publishing: Port Chester, NY, USA, 2004.
22. Aguilar-Luzón, M.d.C.; García-Martínez, J.M.Á.; Calvo-Salguero, A.; Salinas, J.M. Comparative Study between the Theory of Planned Behavior and the Value–Belief–Norm Model Regarding the Environment, on S panish Housewives' Recycling Behavior. *J. Appl. Soc. Psychol.* **2012**, *42*, 2797–2833. [CrossRef]
23. Bamberg, S.; Möser, G. Twenty years after Hines, Hungerford, and Tomera: A new meta-analysis of psycho-social determinants of pro-environmental behaviour. *J. Environ. Psychol.* **2007**, *27*, 14–25. [CrossRef]
24. Farrukh, M.; Raza, A.; Mansoor, A.; Khan, M.S.; Lee, J.W.C. Trends and patterns in pro-environmental behaviour research: A bibliometric review and research agenda. *Benchmarking Int. J.* **2022**. [CrossRef]

25. Onokala, U.; Banwo, A.O.; Okeowo, F.O. Predictors of Pro-Environmental Behavior: A Comparison of University Students in the Untied States and China. *J. Manag. Sustain.* **2018**, *8*, 127. [CrossRef]
26. Aguilar-Luzón, M.C.; Calvo-Salguero, A.; Salinas, J.M. Beliefs and environmental behavior: The moderating effect of emotional intelligence. *Scand. J. Psychol.* **2014**, *55*, 619–629. [CrossRef]
27. Tevdovska, E.S. The impact of emotional intelligence in the context of language learning and teaching. *Seeu Rev.* **2017**, *12*, 125–134. [CrossRef]
28. Yahaya, A.; Bachok, N.S.E.; Yahaya, N.; Boon, Y.; Hashim, S.; Goh, M.L. The impact of emotional intelligence element on academic achievement. *Arch. Des Sci.* **2012**, *65*, 2–17.
29. Zhoc, K.C.; Chung, T.S.; King, R.B. Emotional intelligence (EI) and self-directed learning: Examining their relation and contribution to better student learning outcomes in higher education. *Br. Educ. Res. J.* **2018**, *44*, 982–1004. [CrossRef]
30. Chisholm, C.U. The formation of engineers through the development of Emotional Intelligence and Emotional Competence for global practice. *Glob. J. Eng. Educ.* **2010**, *12*, 6–11.
31. Magano, J.; Silva, C.; Figueiredo, C.; Vitória, A.; Nogueira, T.; Pimenta Dinis, M.A. Generation Z: Fitting Project Management Soft Skills Competencies—A Mixed-Method Approach. *Educ. Sci.* **2020**, *10*, 187. [CrossRef]
32. Tsalaporta, E. Emotional Intelligence for sustainable engineering education: Incorporating soft skills in the capstone chemical engineering capstone design project. In Proceedings of the 10th Engineering Education for Sustainable Development Conference, Cork, Ireland, 14–16 June 2021; pp. 1–8.
33. Campos, C.B.d.; Pol, E. Can the environmental beliefs of workers from environmental certified companies predict their environmental behavior outside the organization? *Estud. De Psicol.* **2010**, *15*, 198–206. [CrossRef]
34. Riemer, M.J. Integrating emotional intelligence into engineering education. *World Trans. Eng. Technol. Educ.* **2003**, *2*, 189–194.
35. United Nations. The 17 Goals. Available online: https://sdgs.un.org/goals (accessed on 8 March 2023).
36. Rego, A.; Fernandes, C. Inteligência emocional: Contributos adicionais para a validação de um instrumento de medida. *Psicologia* **2005**, *19*, 139–167. [CrossRef]
37. Silva, C.; Magano, J.; Figueiredo, C.; Vitória, A.; Nogueira, T. A multi generational approach to project management: Implications for engineering education in a smart world. In Proceedings of the 2020 IEEE Global Engineering Education Conference (EDUCON), Porto, Portugal, 27–30 April 2020; pp. 1139–1148.
38. Kline, R.B. *Principles and Practice of Structural Equation Modeling*; Guilford Publications: New York, NY, USA, 2015.
39. Byrne, B. *Structural Equation Modeling with AMOS: Basic Concepts, Applications, and Programming*, 3rd ed.; Routledge: New York, NY, USA, 2016.
40. Hair, J.; Anderson, R.; Tatham, R.; Black, W. *Multivariate Data Analysis with Readings*; Prentice-Hall: Englewood Cliffs, NJ, USA, 1995.
41. Field, A. *Discovering Statistics Using IBM SPSS Statistics*; Sage: New York, NY, USA, 2013.
42. Fornell, C.; Larcker, D.F. Evaluating structural equation models with unobservable variables and measurement error. *J. Mark. Res.* **1981**, *18*, 39–50. [CrossRef]
43. Cohen, J. *Statistical Power Analysis for the Behavioral Sciences*, 2nd ed.; Lawrence Erlbaum Associates: Hillsdale, NJ, USA, 1988.
44. Hayes, A.F. *Introduction to Mediation, Moderation, and Conditional Process Analysis: A Regression-Based Approach*, 2nd ed.; Guilford Publications: New York, NY, USA, 2017.
45. Nunnally, J.C. An overview of psychological measurement. In *Clinical Diagnosis of Mental Disorders*; Springer: Boston, MA, USA, 1978; pp. 97–146. [CrossRef]
46. Mifsud, M. Environmental education development in Malta: A contextual study of the events that have shaped the development of environmental education in Malta. *J. Teach. Educ. Sustain.* **2012**, *14*, 52–66. [CrossRef]
47. Goleman, D. *Working with Emotional Intelligence*; Bantam Books: New York, NY, USA, 1998.
48. Encinas, J.J.; Chauca, M. Emotional intelligence can make a difference in Engineering Students under the Competency-based Education Model. *Procedia Comput. Sci.* **2020**, *172*, 960–964. [CrossRef]
49. Mitrović Veljković, S.; Nešić, A.; Dudić, B.; Gregus, M.; Delić, M.; Meško, M. Emotional intelligence of engineering students as basis for more successful learning process for industry 4.0. *Mathematics* **2020**, *8*, 1321. [CrossRef]
50. Ranganathan, P.; Aggarwal, R. Study designs: Part 1—An overview and classification. *Perspect. Clin. Res.* **2018**, *9*, 184. [CrossRef] [PubMed]

 *sustainability*

Article

Multidimensional Characteristics of Design-Based Engineering Learning: A Grounded Theory Study

Lina Wei [1,2], Wei Zhang [3,4] and Chenhua Lin [3,4,*]

1   Institute of Medical Education/National Center for Health Professions Education Development, Peking University, Beijing 100191, China
2   School of Education, Peking University, Beijing 100871, China
3   School of Public Affairs, Zhejiang University, Hangzhou 310058, China
4   Institute of China's Science, Technology and Education Policy, Zhejiang University, Hangzhou 310058, China
*   Correspondence: chlin@zju.edu.cn; Tel.: +86-18814851808

Abstract: Design-based engineering learning is an important learning model in the field of engineering education research, and also an important embodiment of sustainable development engineering education. At present, the research is still in the stage of conceptual discussion, and its connotation has not been effectively clarified. In order to construct a theoretical model of design-based engineering learning, this study adopted grounded theory to carry out exploratory research. First, we selected the advanced class of engineering education in Chu Kochen Honors College of Zhejiang University as a follow-up case, conducted interviews with front-line teachers and students, and collected relevant data; second, we adopted the three-level coding technology of open coding-axis coding-selective coding and used NVivo software to extract concept categories from open codes, and established the connection between categories through the axis coding; and finally, multidimensional ideas were developed through core categories, including design practice, interactive reflection, knowledge integration, and circular iteration. The multidimensional conception of design-based engineering learning constructed in this study aims to provide theoretical support for promoting engineering education research. At the same time, it puts forward some useful suggestions on the training of engineering talents for sustainable development in practice.

Keywords: design-based engineering learning; multidimensional conception; grounded theory

Citation: Wei, L.; Zhang, W.; Lin, C. Multidimensional Characteristics of Design-Based Engineering Learning: A Grounded Theory Study. *Sustainability* **2023**, *15*, 3389. https://doi.org/10.3390/su15043389

Academic Editors: Clara Viegas and Natércia Lima

Received: 26 January 2023
Revised: 8 February 2023
Accepted: 10 February 2023
Published: 13 February 2023

1. Introduction and Background

Since the second half of the 20th century, the contradiction between economic development and ecological environmental protection has become increasingly prominent. As a practical activity for humans to explore and transform the world, engineering is regarded as an important means to protect the ecological environment and improve the quality of life. This new generation of engineers has an important mission and responsibility in advancing the global Sustainable Development Goals for engineering education. Without their wisdom and contributions, these goals would not be possible. While leading the reform of international engineering education, the concept of sustainable development education has increasingly become an important choice to improve the quality of engineering education. In order to support the country's long-term innovation capacity and the sustainable development of the social economy, we must promote the innovative development of global engineering education and cultivate a large number of sustainable competitive innovative talents.

Engineering education is intended to train engineers [1,2]. In the field of engineering education, scholars have been concerned about "how people learn engineering" and "the nature of engineering knowledge". Engineering learning is a process of knowledge acquisition, knowledge creation, and ability development that learners develop around the engineering subject field [3–5].

Engineering is a systematic activity of artificial creation, whose essence lies in design [6–8]. In modern engineering practice, design is an important step from overall planning to concrete implementation and a process of technology integration and engineering synthesis [9]. Only design can create a sound engineering framework and make engineering activities form an organic whole [10]. "Engineering design and training is a creative, repetitive and often endless work", according to China's Certification Standards for Engineering Education (Trial). The Board for Certification in Engineering Technology (ABET) states that engineering design is a decision-making process that requires engineers to apply basic science, mathematics, and engineering science to optimize resource conversion and achieve specific goals. In fact, engineering design has rich connotations: at the scheme level, it is the artificial object to achieve the optimal decision combining entity and intention [11,12]. At the process level, it has multidimensional characteristics such as individual reflection, knowledge integration, and social interaction [13,14]. At the activity level, it integrates resources to implement goal-oriented engineering activities [15,16]. For engineering education, design is the core of training engineers [17,18]. Engineers are different from scientists. "Scientists discover the world that has already existed, while engineers create the world that has never existed" [19]. The function of engineers is to solve complex engineering problems in industry, including new products, new processes, and new technologies [2,20]. A teaching and learning process for engineering education that lacks design activities will lead to a lack of students' intersubjectivity, engineering practical ability, and innovative thinking [10].

Influenced by the trend of scientism, the global higher engineering education system has been turning to a "scientific paradigm" since the 1990s [21,22]. Engineering teaching under the scientific paradigm positions the teacher–student relationship as the subject–object relationship, making the focus of engineering education shift to acquiring formal knowledge, and students gradually become "automated" knowledge containers, unable to achieve deep learning, high-level thinking, or conceptual innovation. In the early 20th century, some scholars proposed "design-based learning" (DBL) in the field of education, in which teachers put forward questions to students based on practical problems and adopt a bottom-up approach to enable students to construct and deepen existing knowledge meaningfully while completing design tasks. Furthermore, they constantly learn new knowledge in a repeated cycle and finally obtain the learning mode of products that meet the task requirements [23–26]. Since then, design-based learning has been regarded as an innovative learning mode combined with engineering education practice and has gradually evolved into a unique design-based engineering learning (DBEL) method.

As a powerful inductive learning method, scholars have studied the concept and connotation of design-based engineering learning from multiple perspectives, such as the learning process, teaching means, and learning mode. Some scholars define DBEL as a learning process that is based on authentic engineering design practice and encourages students to construct knowledge in the process of solving engineering problems [23,27,28]. Some scholars define DBEL as a teaching method in which teachers assign challenging tasks to students and create interactive environments so that students can repeatedly memorize and practice the learned knowledge [29,30]. Some scholars believe that DBEL is a brand-new learning mode. Engineering students design specific engineering models by using prior experience and knowledge learned and then modify and redesign the models and schemes in a circular manner to acquire new knowledge in practice [31,32]. Some scholars believe that DBEL is a collaborative optimization engineering learning mode, in which students constantly analyze and design the existing engineering technology system and make improvements in quality, function, cost, price, and other aspects, thus significantly improving the performance of engineering products [30]. Furthermore, some scholars believe that DBEL enables students to design science content in the context of design challenges, thus promoting deep learning [24,33,34]. Based on the existing research and practice, this paper regards "design-based engineering learning" as an extension of the core of "design-based learning" and "engineering learning", and defines it as a learning mode; that is, in the real-engineering problem situation, according to the specific design task, students use their existing knowledge and experience, through several iterations, to create a certain object model to acquire new knowledge and improve their engineering problem-solving ability using a dynamic learning mode. Design-based engineering learning is a learner-centered learning model that helps to highlight the centrality of engineering students in sustainable engineering education.

In the past two decades, design-based engineering learning has been widely promoted in undergraduate engineering education around the world. Famous universities in the United States have reconstructed engineering curriculum systems with design as the main line. For example, the University of Utah has set a "spiral" introduction to engineering courses for freshmen [35], there has been the construction of integrated design courses in the Mechanical Engineering Department of MIT [14], and Purdue University has integrated engineering design into undergraduate and postgraduate education [36]. In China, Professor Gu Peihua, an academic from the Canadian Academy of Engineering, introduced the CDIO engineering learning mode in 2005 and explored and implemented it at Shantou University [37]. Later, the global CDIO initiative cooperation alliance successively attracted Shantou University, Tsinghua University, Yanshan University, Chengdu University of Information Technology, and other universities to join, and successively formed an innovative engineering education model based on the CDIO represented by Shantou University. Since then, many colleges and universities in China have also promoted design-oriented engineering education reform. For example, Xi'an Jiaotong University has set up a basic general core course named "Innovative Thinking and Robot Maker Practice" with design thinking as the main line [38], Chongqing University has carried out a curriculum teaching innovation design based on the BOPPPS teaching mode [39], and Shanghai Jiao Tong University offers the course "Innovative Thinking and Modern Design" based on Kolb's experiential learning cycle theory [40].

Although design-based engineering learning has demonstrated some achievements, the research on design-based engineering learning is still in the stage of connotation exploration, and the question "what are the core characteristics of design-based engineering learning?" has not been well answered. Therefore, the research question to be solved in this study is as follows: what are the core characteristics of design-based engineering learning, and how can the students' learning process be dynamically deduced? In addition, among the three necessary steps of engineering training (engineering knowledge learning, engineering practical experience, and engineering professional training), engineering learning is the study of engineering-related theoretical knowledge, which mainly takes place in undergraduate education [41]. Therefore, in order to solve the above problems, this study longitudinally tracked typical undergraduate engineering education cases, hoping to build a design-based engineering learning feature model, which is of great significance for promoting the construction of sustainable development engineering education theory and practice system.

2. Research Methods and Data Collection

2.1. Grounded Theory

Grounded theory is a type of research method that draws theories from experience from the bottom up, constantly summarizes, compares, connects, and concentrates the collected empirical data, and finally forms theories [42,43]. This study adopted the grounded theory method and collected research data through in-depth interviews and the observation of research objects. Then, the collected research data were coded and analyzed to determine the characteristics of design-based engineering learning and dynamically deduce the design-based engineering learning process model. As an inductive method, grounded theory needs to relate everyday facts and phenomena to theoretical explanations and interpretations, achieving the ability to understand observational relationships and to dynamically explain reality. It focuses on gaining knowledge about the processes behind complex phenomena from qualitative data [44]. The grounded theory combines the flexibility and pragmatism of qualitative methods [45,46]. After 1967, as many authors developed different perspectives [47], the method evolved into different versions, with different terminology and implementation paths. Currently, there are three approaches to grounded theory: inductive framework, interpretive framework, and constructivism framework [48]. In this study, we follow an interpretive framework.

Since the grounded theory was proposed, the academic circle has had a lot of discussion on its research process, among which the "programmed grounded theory" gradually developed around three-level coding and has been widely applied due to its clear process and easy operation [43,49,50]. In this study, we selected "programmed grounded theory" and followed the research process of "defining object–literature discussion–data collection and analysis–establishing preliminary theory–testing theoretical saturation–constructing theory". Using the "open coding–axial coding–selective coding" coding process, the category was gradually abstracted through the analysis of materials, and finally, the core relationship was established to achieve the theoretical construction (see Figure 1) [51].

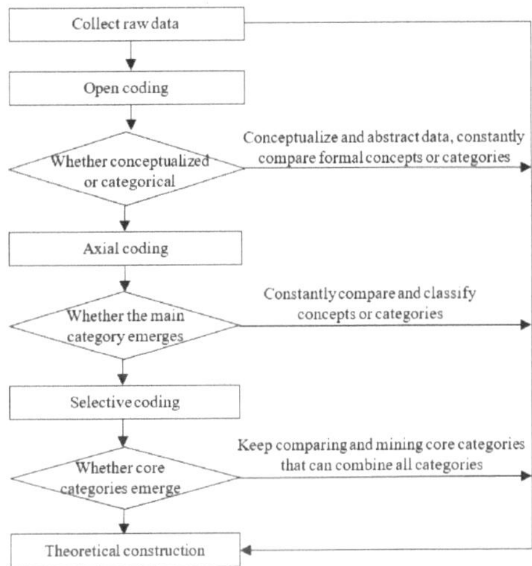

Figure 1. The thinking framework of grounded theory research.

Open coding refers to the process of coding the initial material sentence by sentence, continuously clustering and integrating the key information of the content, and gradually abstracting to form the "conceptual category". The principal axis coding is a process of induction and correlation of related concepts based on open coding, so as to form the "main category". Selective coding is to sort out the relationship and abstract the theory of the spindle coding results and finally form a clear storyline. In the process of operation, we used the software NVivo 12 as the coding tool. NVivo is the most mainstream analysis tool for qualitative research; it supports qualitative research methods and mixed methods. It can collect, organize and analyze interviews, focus group discussions, questionnaires, audio, etc. It is very suitable for analyzing the interview materials for this paper. Through the functions of open coding, node coding, relationship coding, reliability, and validity testing, etc., open coding (through NVivo 12 open coding and node coding) and spindle coding (through relationship coding in NVivo 12) in the research process of the rooted theory were realized. Combined with the literature, actual interviews, and other processes, open coding and spindle coding were constantly abstracted and refined to achieve selective coding and the construction of the final theory.

2.2. Purposeful Sampling

This study adopted purposeful sampling and selected the advanced engineering education class of the Chu Kochen Honors College of Zhejiang University as the tracking case. This class is a pilot reform of engineering education at Zhejiang University. Every year, 40 students are selected from the engineering students in the second semester of the freshman year to prepare separate classes and carry out two-year engineering design courses, aiming at cultivating high-level interdisciplinary talent with innovative abilities. This study conducted in-depth interviews with students and teachers who participated in engineering high school classes and obtained first-hand data. The sample of students included sophomores and juniors who had participated in engineering high school programs, and the sample of teachers included frontline teachers and teaching managers who had participated in engineering high school classes for a long time. Among them, the sample teachers had at least two semesters of engineering design teaching experience and had been unanimously recognized by students and peers in engineering design teaching. Based on this criterion, 20 interviewees were selected, including 7 teachers and 13 students (see Table 1).

Table 1. Basic information table of interview samples.

Basic Information	Characteristics	Teacher	Student	Total
Gender	Male	4	8	12
	Female	3	5	8
Grade	Sophomore year	/	9	9
	Junior year	/	4	4
Number of years of teaching or participating in DBEL	One year	/	9	9
	Two years	/	4	4
	Over two years	7	/	7
Total number		7	13	20

2.3. Interview Design and Implementation

Focusing on the "core characteristics and functional process of design-based engineering learning", the interview outline (see Appendices A and B for the interview outline) was studied and prepared. A semi-structured interview method was adopted, with face-to-face interviews as the main method and telephone interviews as the auxiliary method. Through face-to-face and telephone interviews, 1140 min of interview recordings were obtained. The recording time of teachers' and students' interviews was 431 min and 709 min, respectively. The content of the interviews was transformed into text through "IFLYREC". After collation and modification by two researchers, about 230,000 words of interview manuscripts were finally formed, including 59,000 words for teachers and 172,000 words for students. This laid the foundation for the coding and analysis of the follow-up interview text. In the study, the interviewees were coded as T01–T07 for teachers and S01–S13 for students. See Appendix A for the outline of the teacher interview and Appendix B for the outline of the student interview.

3. Data Coding and Analysis

In this study, qualitative text content analysis and keyword clustering induction methods were used, and NVivo software, a qualitative text analysis tool, was used as the research tool to complete the content analysis of research literature by using its text coding, data visualization, clustering comparison, and other functions. There are two common coding methods: one is to determine the coding node according to the research topic and form a research framework, which is called deductive coding; the second is to code the original text first and integrate it after generating multiple subnodes, that is, inductive coding. In terms of practical operation technology, the layered coding method was adopted, which was divided into three stages: open coding (first-level coding), node coding (second-level coding), and relationship coding (third-level coding) (see Figure 2).

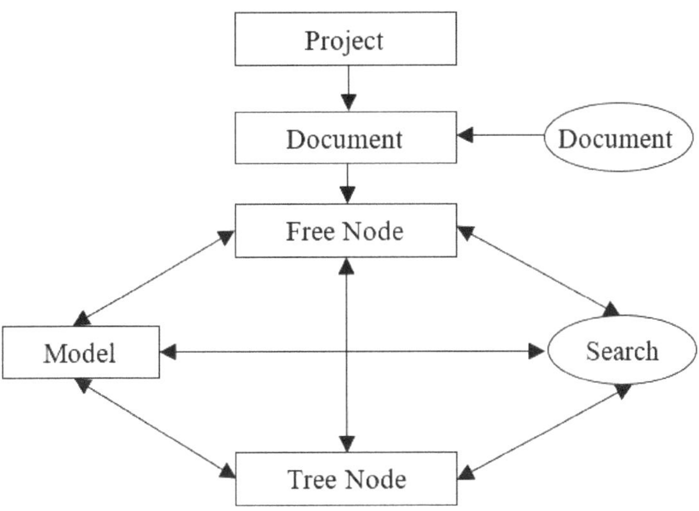

Figure 2. The coding process of NVivo12.

3.1. Open Coding

Using three methods of "line by line coding", "sentence by sentence coding", and "paragraph by paragraph coding", through the process of refining, induction, and comparison, all 20 samples of text were "labeled", 271 initial tags were extracted, and then 271 initial tags were filtered, combined, and classified to form 76 "initial concepts". An example of open coding is shown in Table 2.

Table 2. Examples of DBEL open coding.

Sample Source Data		Initial Concept
S2: I think it's important that you know what to do and what the details are during the design process. Hands-on skills are acquired, and trial and error are crucial.	T3: During the design process, I will give feedback to students as soon as possible to help them adjust and iterate.	aa10 Students' hands-on practice aa67 Trial and error
S9: I like to participate in design projects. Doing projects gives me a sense of accomplishment. When I finish a project, I feel very happy because I can learn a lot.	T6: Our analog integrated circuit and robotics courses are usually held in the first or second year of junior year. At this stage, students have a lot of practical design content, which is relatively difficult.	aa3 Design project
S10: Whenever I have a problem in the learning process, I want to solve it, no matter what the problem is or what the teacher said in class. No matter how many times I try and make mistakes, if I can't solve them, I will keep thinking until I solve them.	T6: The design process is to iterate through the whole process to stimulate students to consider some underlying practical engineering problems.	aa67 Trial and error aa74 The problem is rectified
S12: Making plans with classmates can deepen my impression. When you can explain the whole scheme in its entirety, you will understand the problem very well; If I don't know how to do it, at this time, other students may come to tell me their thoughts and ideas, which can also help me understand what I didn't understand before. This process is very effective.	T1: Engineering students should not only have the professional ability but also have the spirit of teamwork. Integration and innovation are very important for engineering students, and these things need students to practice daily communication and cooperation.	aa41 Student exchange and cooperation

3.2. Axial Coding

In order to further form the main category and subcategory, main axis coding (secondary coding) was carried out in this study, as shown in Table 3. Spindle coding is a more abstract concept, which requires a profound distinction between the relationships and meanings between concepts and categories, thus forming multiple dimensions of the theory. We further abstracted 76 "initial concepts" and finally formed 28 "initial categories", such as real engineering problems, real project content, and challenging tasks. Then, the "initial categories" were classified into 12 "subcategories", and finally, the "subcategories" were further classified and merged into 4 core categories.

Table 3. Axial coding results.

Core Category	Subcategory	Initial Category	Dimension
Design Practice	Challenging task	Real engineering problems, real project content, challenging tasks	Suitable–Not suitable
	Hands-on practice	Students do hands-on experiments; students make the finished product by hand	Attention–Neglect
	Scheme design	Student participation in program design; students use design methods	Suitable–Not suitable
Interactive Reflection	Peer interaction	Group negotiation, class discussion, group cooperation	Abundant–Deficient
	Teacher–student interaction	Teachers interact with students; the teacher gives guidance to the students	Abundant–Deficient
	Social interaction	Physical interaction in the external environment; school–enterprise interaction in foreign cooperation	Abundant–Deficient
Knowledge Integration	Information recognition	Students understand information; students identify external knowledge	Sufficient–Insufficient
	Knowledge correlation	Students identify deficiencies in the design process; students promote knowledge understanding; students relate and integrate knowledge points	Sufficient–Insufficient
	Opinion generation	Students extract knowledge elements; students generate relevant opinions; students consolidate relevant knowledge	Sufficient–Insufficient
Circular Iteration	Trial and error	Students try and make mistakes along the way	Strong–Weak
	Scheme iteration	Redesign based on trial and error results; participate in many design cycles and design scheme iterations	Strong–Weak
	Result from perfection	Students constantly improve the design products; implement the new learning plan according to the adjustment plan	Strong–Weak

3.3. Selective Coding

In the coding process, we constantly consulted the original data and literature, classified and divided the contents of open coding and spindle coding, and then linked the core categories together according to the storyline and used theoretical analysis to condense them into four axis codes, namely design practice, interactive reflection, knowledge integration, and cyclic iteration. Finally, a "storyline" that runs through all materials, categories, and relationships was constructed. Engineering learning based on design emphasizes process rather than results, which is essentially a closed-loop learning mode; that is, engineering students take design practice as a starting point, observe, think from different perspectives, and incorporate them into their own logical system, so as to gain new experiences through constant trial and error and concept transformation. In design-based engineering learning, learners conduct design inquiry based on design projects, further promote learners' self-reflection and development through "interaction" with other learners, teachers, or experts, and finally, obtain systematic knowledge in inquiry action and cycle iteration to achieve knowledge construction. Figure 3 shows the process.

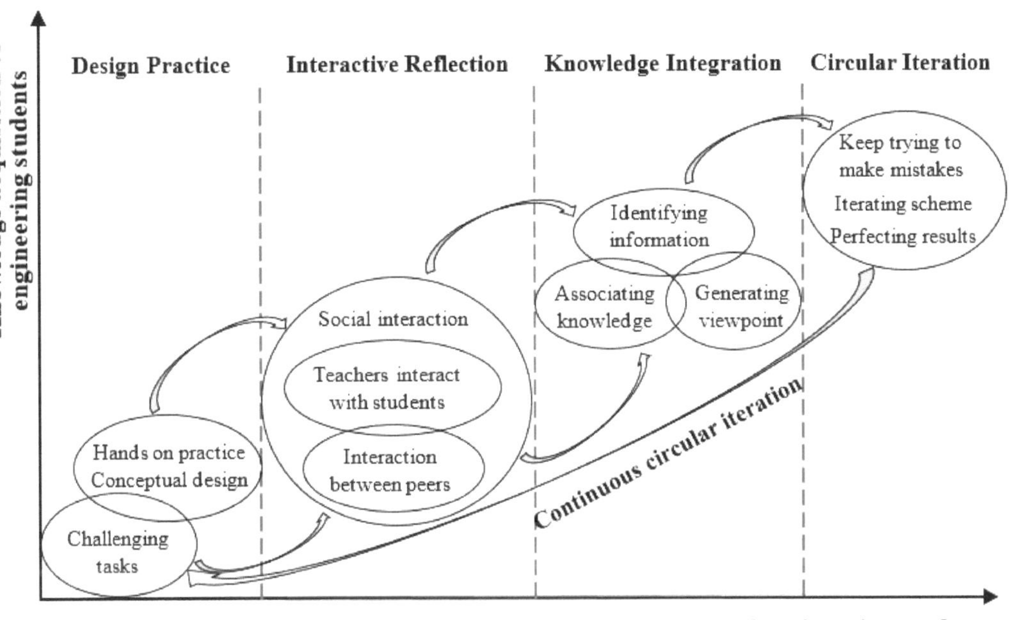

Figure 3. The multidimensional feature model of design-based engineering learning.

3.4. Test of Theoretical Saturation

A theoretical saturation test is a key step to verify whether the interview information is saturated [43]. When new data do not add meaningful contributions to the theory being developed, and no new categories appear [52], it is determined that saturation has been reached. Generally, most of the code can be built and data-saturated in the first 15 interviews [53–56]. In this study, we used the random sampling method to recode the contents of three coded interview materials and found that no new classification concept was formed. In other words, the new document was covered by the previous 43 "initial concepts", proving that it had basically reached theoretical saturation [57]. In addition, NVivo's "coding comparison" function [58] was used in this study to sample the text materials independently coded by two researchers, and the "code consistency percentage" was used to measure the objectivity of the coding. It was concluded that the consistency of 20 original materials in the four dimensions was more than 82.33%, thus ensuring the reliability and validity of the data analysis.

4. Results and Discussion

Through grounded research, this paper refines the multidimensional characteristics of design-based engineering learning and explains the relationship between the four characteristics. In order to further explore learners' understanding and feelings in design-based engineering learning, interview minutes and interview materials of 20 learners were analyzed in detail, and node contents of NVivo were read through the node coding query function of NVivo12. The keywords and key texts in the reference points of design practice, interactive reflection, knowledge integration, and cyclic iteration were analyzed successively.

4.1. Design Practice

Through careful reading and analysis of the node content of the learners' design practice category, it was found that the characteristics of the learners' design practice were mainly manifested as design tasks, process optimization, hands-on practice, etc. (see Figure 4).

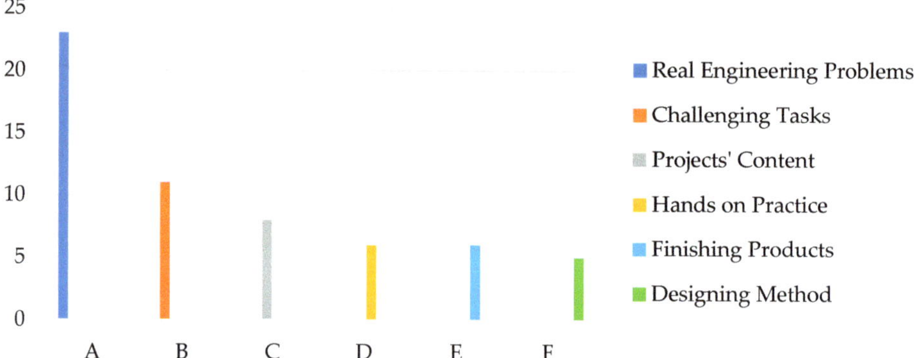

Figure 4. Statistics and analysis of the number of coding nodes in design practice (top 6).

Design practice is an exploratory activity carried out by students around the design situation. It is based on scientific principles or theories, aiming to conceive and implement some practical new products, and the final result is to produce innovative methods to solve problems. The connotation of design practice mainly includes the following aspects: The first is challenging tasks. Learners' learning often requires the support of the environment. In design-based engineering learning, this environment takes the form of challenging and specific engineering design tasks, which can introduce students to the real engineering environment. The second is scheme design. Scheme design is an important way for learners to complete challenging tasks. Learners make use of multidisciplinary knowledge to design, organically integrate learning content with hands-on practice, realize the true meaning of learning, and gradually move from "passive acceptance" to "active exploration". The third is hands-on practice. Engineering learning needs practical "people", who are living individuals with initiative and creativity, rather than machines with only "pure theory". The practicability of the learning subject in the design context is related to the effectiveness of the whole learning process, which reflects the construction of the learning process with "behavior" as the link and "situation" as the bridge.

In interviews, students and teachers clearly expressed the importance of design practice in improving engineering learning outcomes:

Sometimes the teacher tells me a lot of things, and I listen to the fog, then forget. However, for a specific example in the course, if I do it by hand, I will be deeply impressed after the experience, and then I will know how to go further after I find the problem, so practice is very important. (S3)

Engineering design is biased towards practice. Through the design scheme and physical model, I fully practice the engineering knowledge I have learned. (S7)

For engineering students, practical ability should be the most basic. It is very effective to cultivate students' practical ability in design-related practical projects, which can help students simulate the engineering environment. (T03)

Our philosophy is that students in the lower grades should be exposed to robots early. Students will understand the difficult points, key points, and core points of robots in practice. When students have a basic understanding of the whole knowledge point, and then go back to learning machinery, computers, and other related courses, students will know how to combine relevant knowledge to solve engineering problems. (T07)

4.2. Interactive Reflection

It was found that learners' interactive reflection features mainly include group co-operation, peer learning, communication, and discussion. In order to further explore the characteristics of learners' interactive reflection in design-based engineering learning, the number of reference points in learners' interactive reflection was obtained (see Figure 5).

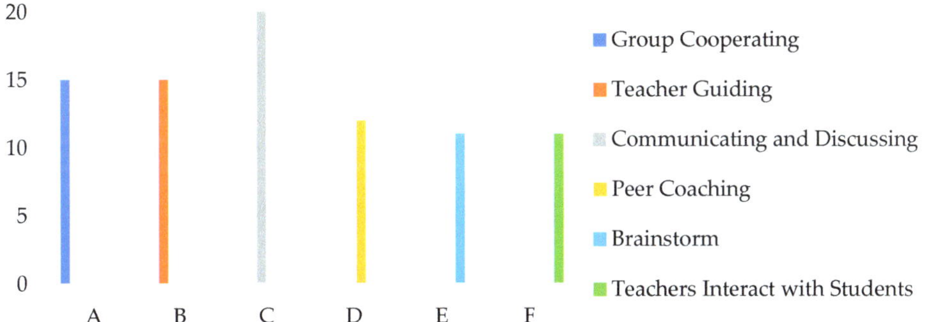

Figure 5. Statistics and analysis of the number of coding nodes in interactive reflection (top 6).

Although many scholars have recognized the role of "reflection" in learning, they all regard reflection as an individual's spontaneous behavior and fail to consider how external stimuli affect the individual's thinking and how reflection affects the behavior of those studied. Through the above coding analysis and theoretical analysis, we know that "design task" and "community interaction" are important forms of design-based engineering learning. The "design task" puts learners into a thinking deadlock where they cannot find a solution to real engineering problems. In this state, individuals' cognitive input and learning efficiency will be very high. At this stage, interactive reflection helps students have an epiphany.

The interaction of design-based engineering learning is mainly manifested in three ways: The first is the interaction between students and students. Discussion and communication among students are effective ways for students to carry out design-based engineering learning, and their own experience is also an important learning context for design-based engineering learning. Design-based engineering learning should make full use of learning resources among students, carry out cooperative learning, and establish a community. In this "community" learning environment, the student's learning process is influenced by teachers' guidance and evaluation, and students' cooperation and sharing. The second is teacher–student interaction. Teacher–student interaction is the interaction between students, teachers, and teaching content. Teachers and students will have a different understanding of the same course content, and they can help improve students' knowledge acquisition and problem-solving abilities by expressing and discussing different viewpoints. This can also further strengthen the emotional communication between teachers and students so that students can obtain positive emotional experiences in the learning process. The third is social interaction. Students are brought together by design activities, and they make full use of engineering drawings and the physical environment of engineering materials in the design process to form a situation of social interaction.

Students can deeply feel the atmosphere of group cooperation in the process of learning. Teachers and students recorded their thoughts on design-based engineering learning in a journal.

The process of completing the design task usually involves the cooperation of students from multiple majors. Students from different majors will be responsible for different modules. In the process of task docking, there must be a lot of work in communication and connection, and interaction is very common. (S9)

The biggest goal of group cooperation is to obtain more satisfactory results after constantly revising the program! (S13)

We don't care about the form of classroom teaching but focus on the content and effect of the lecture, which is the most fundamental. The reality is that students are more likely to take the initiative to learn, be inspired, and gain from cooperation. I think teamwork is very effective. (T05)

4.3. Knowledge Integration

It was found that learners' knowledge integration features mainly included knowledge application, association integration, and association consolidation. In order to further explore the characteristics of learners' knowledge integration in design-based engineering learning, the number of reference points in learners' knowledge integration was obtained through research and analysis (see Figure 6).

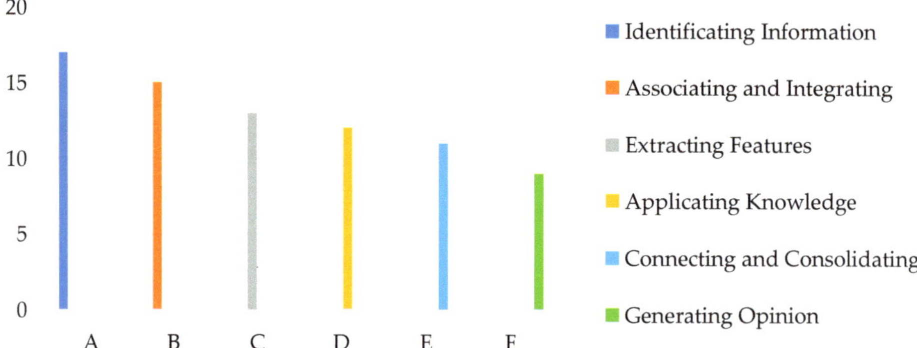

Figure 6. Statistics and analysis of the number of coding nodes in knowledge integration (top 6).

The purpose of design-based engineering learning is to help learners identify real engineering problems and apply knowledge to solve them. The knowledge integration stage is an important stage to promote learners' learning and produce behavioral changes. In this stage, students can not only extract feasible ideas but also combine these ideas with what they have learned and make plans for the next step with new actions, thus promoting changes in their behavior patterns. Therefore, design-based engineering learning has strong interdisciplinarity, and its interdisciplinary integration is equally important to engineering design. Knowledge integration in design-based engineering learning is mainly manifested in the following two ways: The first is knowledge association and integration. Design-based engineering learning is complex and integrated, and students need to fully integrate engineering-related knowledge. The second is the integration of process and result. Design-based engineering learning is not a single static process, but a comprehensive dynamic learning process that advocates learning for the purpose of problem-solving and cultivates a series of high-level abilities of learners.

The teachers and students in the interview reflected on the integration and application of knowledge:

I was deeply impressed when I finished the big design assignment. One of my classmates directly made a small program of an artificial intelligence volleyball team, in which two volleyball teams designed by him could play together. This program requires a lot of knowledge, and I admire it. It seems that to make a great thing, one must have strong knowledge integration ability, this requirement is very high. (S1)

I think knowledge integration is crucial in the realization of engineering design. The design process is not monolithic, but multidisciplinary, and you need to constantly evaluate how students are applying their knowledge, and provide appropriate guidance based on feedback. (T3)

4.4. Circular Iteration

It was found that learners' cyclic and iterative characteristics mainly showed continuous trial and error, scheme adjustment, correlation integration, and so on. In order to further explore the characteristics of learners' cyclic iteration in design-based engineering learning, the number of reference points in learners' cyclic iteration was obtained through research and analysis (see Figure 7).

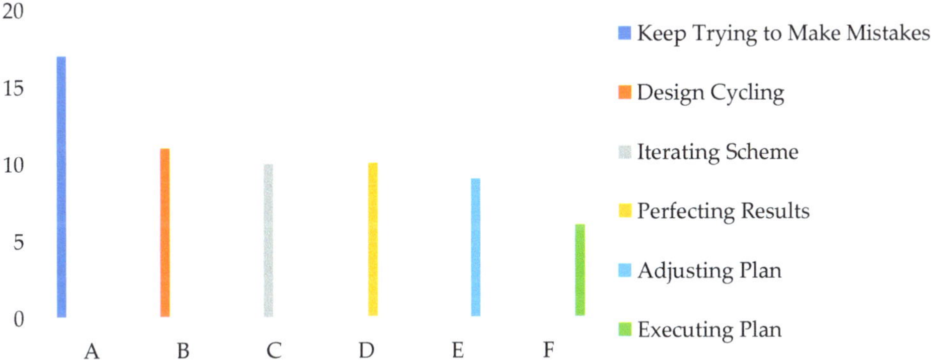

Figure 7. Statistics and analysis of the number of coding nodes in circular iteration (top 6).

The core of engineering is design, and the key to design is iteration. In design-based engineering learning, learners need to constantly try and make mistakes, take active actions, diagnose and debug the information generated by actions, and thus promote their own deep learning. In this process, learners deduce possible interpretations (logical reasoning), anticipate some results (reasoning), or summarize (generalization) existing information from other information, which is exactly the learning process based on engineering problems.

On the whole, in the learning context of design-based engineering learning, the outside world constantly gives feedback on the learning process of students, and students constantly reflect on their own design methods and promote the absorption of implicit thinking by constantly clarifying the underlying thinking process, basic principles, and progress.

In the interview minutes, there were more than 30 descriptions of timely feedback from the course teachers, teaching assistants, and peers, such as the following:

Engineering and design is a process of trial and error. Experience has taught me that trial and error have a cost. If I obtain results after some tries, then I will be happy. If I keep trying and making errors without finding a solution, I find the process very boring. (S7)

Sometimes I get caught up in self-inspired learner mode, where I've spent a lot of time but the feedback isn't satisfactory. There is definitely a problem with this process. You may think you have learned it, but you may not have understood it at all. At this time, we need to reflect on our own practice process and then communicate with classmates to understand the reasons, and finally establish a connection with the knowledge system, or apply it to practical problems. (S1)

> *Design-oriented curriculum teaching emphasizes a closed loop in teachers' instructional design before, during, and after class, as well as a self-cycle in students' individual learning. (T6)*

4.5. Operational Definition

Based on the grounded theory, this paper proposes that design practice, interactive reflection, knowledge integration, and circular iteration should be dynamically deduced in the process of DBEL. Students can realize personal knowledge construction and professional ability improvement by connecting old experiences, engaging cognition, summarizing reflection, and other cognitive cycles. In the early coding, we identified key nodes of DBEL features, and we made operational definitions for each feature construct through detailed analysis of key behavior examples in the learning process of the interview cases, as shown in Table 4.

Table 4. Operational definition of DBEL feature constructs.

Characteristic Dimension	Operational Definition
Design Practice	Engineering students participate in real design projects and hands-on solutions to real engineering problems.
Interactive Reflection	Engineering students interact with people and products in the process of design practice and seek new perspectives to solve design problems.
Knowledge Integration	Engineering students receive external information, integrate internal knowledge, and transform knowledge and information into problem-solving strategies.
Circular Iteration	In the process of design, engineering students constantly apply trial and error, iterate, verify the design scheme, and finally, solve the problem.

When sorting the interview data, the researchers selected 67 coding units from the interview data that were in line with the characteristics of DBEL and expressed them in a relatively independent and complete form. The selected codes follow the classification exclusion principle, where each encoding unit has only one classification. If a particular coding unit is classified as "other characteristics", the reason should be stated. Finally, this study formed the feature coding table of each dimension shown in Table 5.

Table 5. Operational coding of DBEL feature constructs.

Characteristic Dimension	Operational Coding
Design Practice	Engineering students design experiments and optimize the design process and methods
	Engineering students learn through real projects and challenging assignments
	Engineering students carry out hands-on experiments and make finished products Engineering students apply technology and tools to design engineering problem solutions
Interactive Reflection	Engineering students integrate internal information to generate new ideas
	Engineering students identify external perspectives and integrate them
	Engineering students use multidisciplinary approaches and tools to solve engineering problems
	Engineering students integrate theoretical knowledge with practical experience

Table 5. *Cont.*

Characteristic Dimension	Operational Coding
Knowledge Integration	Engineering students participate in group cooperation and group negotiation
	Engineering students reflect on the learning process in collaboration
	Engineering students share and exchange ideas and generate new ideas
	Engineering students share solutions with their classmates
Circular iteration	Engineering students are constantly experimenting with trial and error, trying new solutions
	Engineering students identify the deficiencies in the design process and constantly improve and practice
	Engineering students seek the best solution to improve the learning effect

5. Conclusions

5.1. Theoretical Contributions

This study deconstructed the core features of design-based engineering learning through grounded theory, proposed a new concept of design-based engineering learning, and constructed a circular learning circle that integrates design practice, interactive reflection, knowledge integration, and circular iteration. It was found that design-based engineering learning is a new concept that starts with the process of seeking solutions to ill-structured problems in engineering learning under uncertain conditions. In essence, DBEL is a process in which engineering students constantly practice and construct the design task, thus promoting the continuous improvement of their professional ability. In this study, design-based engineering learning's operational definition and operational coding have laid a foundation for the subsequent measurement of DBEL and testing of DBEL learning effects. The exploration of feature conception in this study contributes to the theoretical development of design-based engineering learning and is of great significance for promoting the construction of higher engineering education system based on the concept of sustainable development.

5.2. Practical Enlightenment

This study provides practical enlightenment for engineering students to learn. Due to the lag effect of practical feedback, engineering students' personal behaviors in the learning process will have a certain degree of blindness. Most students' learning behaviors stay in a certain stage of learning, but cannot reach the cycle and iteration phase. The multidimensional characteristics proposed in this study provide methodological guidance for engineering students to participate in engineering learning. Based on this, on the one hand, engineering students can continuously accumulate relevant knowledge and ability through design practice and iterative attempts and constantly improve their ability to solve engineering problems. On the other hand, continuous reflection can broaden new horizons and optimize problem solutions through knowledge integration.

At present, the field of engineering education has begun to try to promote the deep integration of the sustainable development concept and engineering education through multidisciplinary integrated engineering project design. The situational nature of design-based engineering learning makes it possible to integrate cognitive skills and professional knowledge into specific situations. To carry out the study practice of design-based sustainable development engineering, firstly, an integrated knowledge framework containing professional knowledge and sustainable development content should be provided, and modular training should be implemented for engineering students to realize the visualization of knowledge content related to sustainable development. Secondly, specific problem scenarios should be provided; for example, civil engineering students could be provided with a study project on the construction of a hydropower station, and the students could be encouraged to consider various issues from a sustainable perspective during the project design process, including the biodiversity around the hydropower station, resident relocation, transportation, etc. Thirdly, it provides the "scaffolding" to understand and become familiar with the requirements of sustainable development engineering. Software simulation, video learning, and group communication are used to help students grasp the rich connotation of sustainable development engineering education and meet specific sustainable development goals in real engineering design and practical project operation. Finally, it examines engineering design activities from the perspective of constructivism, focuses on the process of establishing the connection between individuals and society with engineering knowledge of sustainable development, promotes the interaction between teachers and students in engineering design and engineering products, and promotes cooperative learning and open sharing in the process of design activities.

5.3. Limitations and Prospects

As an exploratory study, this study still has some shortcomings. For example, since the coding materials in this study are mainly interview records, supplemented by a small number of internal materials, the coverage of interviewees and the richness of primary data have limitations in the coding conclusions. In view of this, future research needs to expand the scope and number of interviewees, and further improve the interview outline in multiple rounds of interviews, so that the research can show the implementation of design-based engineering learning to the greatest extent and scope. Future studies need to further explore the influence mechanism of different dimensions on engineering students' learning performance.

Author Contributions: Conceptualization, L.W. and W.Z.; data collection, methodology, L.W. and C.L.; writing—original draft preparation, L.W.; writing—review and editing, C.L. and W.Z. All authors have read and agreed to the published version of the manuscript.

Funding: This paper is supported by the National Natural Science Foundation of China program "Mechanism and Effectiveness of Design-based Engineering Learning" (72074191), the Zhejiang Provincial Natural Science Foundation program, "Research on the mechanism of engineering learning and its intervention strategy based on community of practice" (LZ22G030004), and the Chinese Society of Academic Degree and Graduate Education program, "Research on the Cultivation of Practical Ability of Chinese Professional Degree Postgraduates Facing the Demand of Strategic Emerging Industries" (2020ZD1014).

Institutional Review Board Statement: Not applicable.

Informed Consent Statement: Not applicable.

Data Availability Statement: Not applicable.

Conflicts of Interest: The authors declare no conflict of interest.

Appendix A

Semi-structured interview questions for the teachers:

(1) What do you think is the position of design-based engineering learning in engineering education?

(2) From your teaching and management experience, what principles should be followed in the implementation of design-based engineering teaching, and what kind of learning environment should be created?

(3) Based on your teaching and management experience, can design-based engineering learning improve the learning outcomes of engineering students, and what is the key to improving their performance?

(4) According to your teaching and management experience, what difficulties have you encountered in teaching engineering design courses? How was it solved?

(5) What are the key elements to the success of design-based engineering learning besides the efforts of teachers?

(6) What do you think is the development stage of design-based engineering learning in China compared with foreign countries? What foreign countries are worth our learning and referencing?

(7) What are your suggestions for the development of design-based engineering learning in China?

(8) How does the Teaching Steering Committee of Chinese colleges and universities promote the development of design-based engineering learning?

Appendix B

Semi-structured interview questions for the students:

(1) Have you ever participated in design-based engineering studies during your four years at university?

(2) Can you introduce the situation of the learning project that you participated in (participation mode and period, course content, teaching method, organization method, the evaluation method of student learning outcomes, etc.)?

(3) What do you think are the main points and characteristics of design-based engineering learning? Can you describe your learning process, the key stages and key events in your learning of design-based engineering?

(4) What problems and challenges have you experienced in the learning process? How did you solve it? Could you share with me in detail the process of solving this problem?

(5) How did you communicate with team members during the learning process of design-based engineering? What benefited most from communication with team members?

(6) Do you think the practice of design-based engineering learning in your current major can meet your needs for employment or further study in the future? If not, what are the problems? How can it be improved?

(7) Are you satisfied with the supply of various teaching resources and organizational management provided by colleges and universities? If not, what are your suggestions?

References

1. Wang, P.; Gu, J.; Liu, W. *Foundation of Engineering Education*; Higher Education Press: Beijing, China, 2015.
2. Xiang, C. Engineering Learning: Model Change and Theoretical Interpretation. *Res. High. Educ. Eng.* **2015**, *4*, 55–63.
3. Stevens, R.; O'Connor, K.; Garrison, L.; Jocuns, A.; Amos, D.M. Becoming an Engineer: Toward a Three Dimensional View of Engineering Learning. *J. Eng. Educ.* **2008**, *97*, 355–368. [CrossRef]
4. Litzinger, T.; Lattuca, L.R.; Hadgraft, R.; Newstetter, W. Engineering Education and the Development of Expertise. *J. Eng. Educ.* **2011**, *100*, 123–150. [CrossRef]
5. Zheng, J.; Xing, W.; Zhu, G.; Chen, G.; Zhao, H.; Xie, C. Profiling self-regulation behaviors in STEM learning of engineering design. *Comput. Educ.* **2019**, *143*, 103669. [CrossRef]
6. Wallace, K. *Educating Engineers in Design: Lessons Learnt from the Visiting Professor's Scheme*; The Royal Academy of Engineering: London, UK, 2005.
7. Stanton, N.A.; Salmon, P.M.; Rafferty, L.A.; Walker, G.H.; Baber, C.; Jenkins, D.P. *Human Factors Methods: A Practical Guide for Engineering and Design*; CRC Press: Boca Raton, FL, USA, 2017.
8. Cross, N. *Engineering Design Methods: Strategies for Product Design*; John Wiley & Sons, Harper & Row: New York, NY, USA, 2000.
9. Yin, R.Y.; Wang, Y.L.; Li, B.C. *Engineering Philosophy*; Higher Education Press: Beijing, China, 2013; pp. 174–175.
10. Caws, P.; Mitcham, C. *Thinking through Technology: The Path between Engineering and Philosophy*; The University of Chicago Press: Chicago, IL, USA, 1994.
11. Naomasa, N. *Automatic Design*; Marushan: Tokyo, Japan, 1971.
12. Hershauer, R.B.J.C. The new science of management decision, revised edition; Review by: Henry Mintzberg. *Aca Manag. Rev.* **1977**, *3*, 161–162.
13. Kimbler, D.L.; Watford, B.A.; Davis, R.P. Symbolic modelling and design methodology. *Des. Stud.* **1988**, *9*, 208–213. [CrossRef]
14. Sheppard, S.; Jenison, R. Examples of freshman design education. *Int. J. Eng. Educ.* **1997**, *13*, 248–261.
15. Benton, M.; Seireg, A. Factors Influencing Instability and Resonances in Geared Systems. *J. Mech. Des.* **1981**, *103*, 372–378. [CrossRef]

16. Willem, R.A. On knowing design. *Des. Stud.* **1988**, *9*, 223–228. [CrossRef]
17. National Academy of Engineering (NAE). *The Engineer of 2020*; The National Academies Press: Washington, DC, USA, 2004.
18. Miranda, J.; Navarrete, C.; Noguez, J.; Molina-Espinosa, J.-M.; Ramírez-Montoya, M.-S.; Navarro-Tuch, S.A.; Bustamante-Bello, M.-R.; Rosas-Fernández, J.-B.; Molina, A. The core components of education 4.0 in higher education: Three case studies in engineering education. *Comput. Electr. Eng.* **2021**, *93*, 107278. [CrossRef]
19. Li, P.G.; Xu, X.D.; Chen, G.S. Research on problems and Reasons of Chinese undergraduate engineering education practice. *Adv. Mater. Res.* **2012**, *433–440*, 1535–1539.
20. Osgood, L.; Johnston, C.R. Design and Engineering: A Classification and Commentary. *Educ. Sci.* **2022**, *12*, 232. [CrossRef]
21. Seely, B.E. Patterns in the history of engineering education reform: A brief essay. In *Educating the Engineer of 2020: Adapting Engineering Education to the New Century*; National Academy of Science: Washington, DC, USA, 2005; pp. 114–130.
22. Rolstadås, A.; Browne, J. *New Challenges in Manufacturing Engineering Education*; Recent Advances in Mechanical Engineering; Springer: Singapore, 2023; pp. 45–54.
23. Goel, A.K.; Garza, A.G.D.S.; Grué, N.; Murdock, J.W.; Recker, M.M.; Govindaraj, T. Towards design learning environments—I: Exploring how devices work. In *Intelligent Tutoring Systems*; Springer: Berlin/Heidelberg, Germany, 1996; pp. 493–501. [CrossRef]
24. Kolodner, J.L. Facilitating the Learning of Design Practices: Lessons Learned from an Inquiry into Science Education. *J. Ind. Teach. Educ.* **2002**, *39*, 9–40.
25. Mehalik, M.M.; Schunn, C. What Constitutes Good Design? A Review of Empirical Studies of Design Processes. *Int. J. Eng. Educ.* **2006**.
26. Nelson, C.E. Critical thinking and collaborative learning. *New Dir. Teach. Learn.* **1994**, *1994*, 45–58. [CrossRef]
27. Hadgraft, R.G.; Kolmos, A. Emerging learning environments in engineering education. *Australas. J. Eng. Educ.* **2020**, *25*, 3–16. [CrossRef]
28. Dym, C.L.; Shames, I.H. *Introduction to the Calculus of Variations*; Chapman & Hall: New York, NY, USA, 2013.
29. Polishetty, A.; Chandrasekaran, S.; Goldberg, M.; Littlefair, G.; Steinwedel, J.; Stojcevski, A. Enhancing student learning outcomes in manufacturing engineering through design based learning. In Proceedings of the Australian Association of Engineering Education 2014 Conference, Wellington, New Zealand, 8–12 December 2014. [CrossRef]
30. Johri, A.; Olds, B.M. Situated Engineering Learning: Bridging Engineering Education Research and the Learning Sciences. *J. Eng. Educ.* **2011**, *100*, 151–185. [CrossRef]
31. Kimmel, S.J.; Deek, F.P.; Kimmel, H.S. Using a Problem-Solving Heuristic to Teach Engineering Graphics. *Int. J. Mech. Eng. Educ.* **2004**, *32*, 135–146. [CrossRef]
32. Lee, C.-S.; Su, J.-H.; Lin, K.-E.; Chang, J.-H.; Lin, G.-H. A Project-Based Laboratory for Learning Embedded System Design With Industry Support. *IEEE Trans. Educ.* **2009**, *53*, 173–181. [CrossRef]
33. Tempelman, E.; Pilot, A. Strengthening the link between theory and practice in teaching design engineering: An empirical study on a new approach. *Int. J. Technol. Des. Educ.* **2010**, *21*, 261–275. [CrossRef]
34. Xiang, C. *Research on Design-Based Engineering Learning*; South China University of Technology Press: Guangzhou, China, 2016.
35. Starkey, J.M.; Midha, A.; DeWitt, D.P.; Fox, R.W. Experiences in the integration of design across the mechanical engineering curriculum. In Proceedings of the 1994 IEEE Frontiers in Education Conference, San Jose, CA, USA, 2–6 November 1994; pp. 464–468. [CrossRef]
36. Little, P.; Cardenas, M. Use of Studio Methods In The Introductory Engineering Design Curriculum. *J. Eng. Educ.* **2020**, *90*, 309–318. [CrossRef]
37. Gu, P.; Hu, W.; Lu, X.; Bao, N.; Lin, P. From CDIO in China to CDIO in China: Development path, Influence and reasons. *Res. High. Educ. Eng.* **2017**, *1*, 24–43.
38. Li, H.; Zhu, A.; Chen, T.; Zhang, Y. Research on Teaching Practice of engineering design ability Cultivation based on Design thinking. *Res. High. Educ. Eng.* **2022**, *3*, 85–90.
39. Jin, X.; Li, L.; Du, J.; Qiu, Y. Teaching Innovation Design Based on BOPPPS Model—A Case study of "Mechanical Design" course. *Res. High. Educ. Eng.* **2022**, *6*, 19–24.
40. Zhang, Z.; Zhang, G.; Zhu, J. Innovation design ability Cultivation based on Kolb experiential Learning Cycle. *Res. High. Educ. Eng.* **2020**, *1*, 74–78, 136.
41. Li, P.; Xu, X.; Chen, G. Analysis of problems and causes of practical teaching of undergraduate engineering education in China. *Res. High. Educ. Eng.* **2012**, *3*, 1–6.
42. Strauss, A.; Corbin, J. *Grounded Theory Methodology: An Overview Handbook of Qualitative Research*; Sage Publications: New York, NY, USA, 1994.
43. Glaser, B.G.; Strauss, A.L. *The Discovery of Grounded Theory: Strategies for Qualitative Research*; Routledge: New York, NY, USA, 1967.
44. Wuelser, G.; Pohl, C. How researchers frame scientific contributions to sustainable development: A typology based on grounded theory. *Sustain. Sci.* **2016**, *11*, 789–800. [CrossRef] [PubMed]
45. McDonald, S. Studying actions in context: A qualitative shadowing method for organizational research. *Qual. Res.* **2005**, *5*, 455–473. [CrossRef]
46. Kelemen, M.; Rumens, N. Pragmatism and heterodoxy in organization research: Going beyond the quantitative/qualitative divide. *Int. J. Organ. Anal.* **2012**, *20*, 5–12. [CrossRef]
47. Goulding, C. Grounded theory: The missing methodology on the interpretivist agenda. *Qual. Mark. Res. Int. J.* **1998**, *1*, 50–57. [CrossRef]

48. Sebastian, K. Distinguishing between the Strains Grounded Theory: Classical, Interpretive and Constructivist. *J. Soc.* **2019**, *3*, 1–9.
49. Strauss, A.L.; Corbin, J.M. *Basics of Qualitative Research: Techniques and Procedures for Developing Grounded Theory*; Sage Publications: Thousand Oaks, CA, USA, 1998.
50. Charmaz, K.C. *Constructing Grounded Theory: A Practical Guide Through Qualitative Analysis*; Pine Forge Press: Thousand Oaks, CA, USA, 2006.
51. Strauss, A.L. *Qualitive Analysis for Social Scientists*; Cambridge University Press: Cambridge, UK, 1987.
52. Ligita, T.; Harvey, N.; Wicking, K.; Nurjannah, I.; Francis, K. A Practical Example of Using Theoretical Sampling throughout a Grounded Theory Study: A Methodological Paper. *Qual. Res. J.* **2019**, *20*, 116–126. [CrossRef]
53. Guest, G.; Bunce, A.; Johnson, L. How Many Interviews Are Enough? An Experiment with Data Saturation and Variability. *Field Methods* **2006**, *18*, 59–82. [CrossRef]
54. Thomas, S.P.; Pollio, H.R. Listening to patients: A phenomenological approach to nursing research and practice. *J. Adv. Nurs.* **2010**, *42*, 539.
55. Creswell, J.W. *Qualitative Inquiry and Research Design: Choosing among Five Traditions*; Sage Publications: Thousand Oaks, CA, USA, 1998.
56. Boyd, C.O. Philosophical foundations of qualitative research. In *Nursing Research: A Qualitative Perspective*; Jones & Bartlett Learning: Burlington, MA, USA, 1993; pp. 66–93.
57. Saunders, B.; Sim, J.; Kingstone, T.; Baker, S.; Waterfield, J.; Bartlam, B.; Burroughs, H.; Jinks, C. Saturation in qualitative research: Exploring its conceptualization and operationalization. *Qual. Quant.* **2018**, *52*, 1893–1907. [CrossRef]
58. Edhlund, B.; McDougall, A. *NVivo 12 Essentials*; Lulu Press: Morrisville, NC, USA, 2018.

MDPI AG
Grosspeteranlage 5
4052 Basel
Switzerland
Tel.: +41 61 683 77 34

Sustainability Editorial Office
E-mail: sustainability@mdpi.com
www.mdpi.com/journal/sustainability